GLOBAL HEALTH

Agenda Human Geographies

The Agenda Human Geographies series explores contemporary issues, debates and controversies from geographical perspectives and presents the distinctive contributions offered by the discipline. It examines a wide range of topics such as health, globalization, development, justice, migration, race, gender and sexuality, inequality, culture and crime. New and emerging areas as well as established fields come under the spotlight in thought-provoking and incisive discussions by leading scholars. Books are written for an international readership studying, researching and engaging with geographical thought. They are suitable for upper-level undergraduate and postgraduate course use as well as scholars working in geography and related disciplines.

Published

Global Health: Geographical Connections
Anthony C. Gatrell

GLOBAL HEALTH
Geographical Connections

ANTHONY C. GATRELL

agenda
publishing

© Anthony C. Gatrell 2023

This book is copyright under the Berne Convention.
No reproduction without permission.
All rights reserved.

First published in 2023 by Agenda Publishing

Agenda Publishing Limited
PO Box 185
Newcastle upon Tyne
NE20 2DH
www.agendapub.com

ISBN 978-1-78821-499-5 (hardcover)
ISBN 978-1-78821-500-8 (paperback)

British Library Cataloguing-in-Publication Data
A catalogue record for this book is available from the British Library

Typeset by Newgen Publishing UK
Printed and bound in the UK by CPI Group (UK) Ltd, Croydon, CR0 4YY

CONTENTS

Preface vii
List of figures, tables and boxes xi

1 Introduction 1

2 Unequal health I: determinants and regional examples 19

3 Unequal health II: key themes 45

4 Governing global health 73

5 People on the move: the dispossessed and their health and wellbeing 93

6 Materials on the move: out of the ground, and across the globe 115

7 Airs, waters and places 131

8 Infections on the move 151

9 Climate change and global health 175

10 Conclusions: global health and cross-cutting themes 189

References 199
Index 231

PREFACE

Although I am a geographer who has been interested in health issues for 40 years, my research has been almost exclusively focused on countries in the Global North, and even more exclusively on the United Kingdom. More specifically, it has embraced geographical epidemiology – the study of disease and illness in particular places – involving a mix of (usually) quantitative and (sometimes) qualitative methods.

The discipline of health geography, which Susan Elliott and I have sought to summarize in three editions of *Geographies of Health* (2015), has undergone various shifts in emphasis over the last 30 years or so. I have been led to write this present book in part because of the relative lack of research by geographers on global health. Much of the recent focus in health geography has been on topics that have (for example) explored the health and wellbeing of those living near green spaces and attractive walking environments. Yet, in many parts of the world, walking is less for pleasure and more for access to clean water or escape from violence. I hope here to have addressed key issues that affect those living and working outside Anglo-America and Europe; for sure, I have cast my net wide and have endeavoured to cite literature written by those living and working in regions other than my own.

The subject of global health demands an interdisciplinary perspective. The late Paul Farmer and his colleagues have cited anthropology, sociology, political economy and history as the key disciplines (Farmer *et al.* 2013). In this book I am adding geography to the mix, recognizing of course that it too draws theory and concepts from those and other disciplines. As Herrick and Reubi suggest in the introduction to their edited collection (2017b: x), "Geography has not yet carved out a disciplinary niche within the diffuse domain that constitutes global health". I wish to help the inscription, which remains insufficiently "carved": more than ten years have passed since Brown and Moon's (2012) important commentary on geography and global health, although the brief chapter by Brown and Taylor (2018) in Crooks, Andrews and Pearce's comprehensive *Handbook of Health Geography* offers a tantalizing taste of a research agenda and other chapters in that collection discuss global health concerns. So too does a chapter on "global health geographies" in the excellent co-authored text by Brown and colleagues (2018). Clare Herrick is doing much to bring geographers to the table, as my references to her work make clear. However, scrutiny of recent issues of the subdiscipline's leading journal, *Health & Place*, suggests that, despite a growing number of papers, the topic has yet to really fire the imagination of the current generation of health geographers.

It seems to be the case, certainly in the Covid-19 era and for lay audiences, that global health for some means largely a focus on pandemics. I want to show that it embraces so much more than this. There is a wealth of literature on the various topics I cover, drawn (as the references make clear) from a wide variety of disciplines and by authors from the Global South as well as the North. I have always enjoyed ranging across disciplinary boundaries, and I hope the book may therefore appeal to those working outside my "home" discipline of geography. My debt to medical anthropologists in particular will become clear to readers.

This book is subtitled *Geographical Connections*. This indicates that I wish not only to draw out the ways in which geographers can contribute to, or *connect with*, the global health agenda but also, more importantly, to suggest that we live in a *connected* world in which what matters is the sets of relations between elements or entities, whether human or material, social or organizational – or, indeed, all of these. Global health means focusing on issues that impact on particular countries but, crucially, span international borders. As geographers looking at global health, we are concerned with describing and explaining place-to-place variation and connectivities in health (and disease), for which the explanations are multifaceted and demand an engagement with politics, economics and environmental science. But it also means a concern with inequality. At the same time, it is important not to overplay a litany of health problems, illness and disease. Although the contents suggest that the book might well have been titled *Global Disease*, it is clear that people and communities can resist being labelled (even stigmatized) as universal victims of political and economic forces beyond their control.

Aside from those on whose research I have drawn, many others have helped me, particularly those showing personal interest and support in the project. I am especially grateful to Camilla Erskine at Agenda Publishing for her keen interest in the project and her encouragement throughout. I also thank Mike Richardson for his very careful copy-editing of the manuscript. Two reviewers, both of the original proposal and the submitted manuscript, made very helpful and detailed comments, which I have sought to incorporate. I thank them sincerely.

The references section indicates where many of my intellectual debts are owed, but the interest in this project shown by friends and family has meant a great deal. My twin brother Peter commented on the proposal, and anyone interested in refugee history will be familiar with his work, notably his *The Unsettling of Europe* (2019). My wife Caroline, and daughters Anna and Emma, have followed its progress keenly and have also offered considerable support. Their love and inspiration – as scholars themselves – has, for a writer on global health, meant the world.

Tony Gatrell
Lancaster

LIST OF FIGURES, TABLES AND BOXES

Figures

1.1	Haiti as a "living laboratory for the study of affliction"	4
1.2	The "Brandt line" separating the Global North ("developed") from the Global South ("developing")	5
1.3	Musgamagw Tsawataineuk protests at a salmon fish farm	7
1.4	Space–time convergence	11
2.1	Institutional birth delivery coverage according to wealth quintiles and place of residence in selected countries	28
2.2	Brazil: regions and states	30
2.3	Poverty, illiteracy and life expectancy in Brazil	32
2.4	India: states and territories	34
2.5	China: provinces	35
2.6	Life expectancy at birth in China, 2010	36
2.7	Provinces of Mozambique	38
2.8	Provinces of the Democratic Republic of the Congo	41
3.1	Poster of Indian actor Jackie Shroff promoting cigarettes	47
3.2	Dharavi slum in Mumbai, India	56
3.3	Contrasting places: the Paraisópolis favela and luxury buildings in São Paulo, Brazil	57
3.4	Age-standardized disability-adjusted life years resulting from depression	59
3.5	Palestinian children in the Gaza Strip	63
3.6	Association between sustainable development and subjective wellbeing	65

3.7	Association between child mortality and provision of physicians, 2002	68
4.1	World Health Organization HQ, Geneva, Switzerland	76
4.2	GAVI: Global Alliance for Vaccines	82
5.1	Number of internally displaced persons, December 2020	96
5.2	Mapping deaths of people trying to cross the English Channel, 2019–21	99
5.3	Balukhali Rohingya refugee camp in Ukhia, Cox's Bazar, Bangladesh, February 2019	102
5.4	The "Calais jungle", October 2016	103
6.1	Extracting e-waste at the Agbogbloshie scrap market, Accra, Ghana	125
6.2	"Living" in Yemen: the devastation of war	127
7.1	Aerial view of the industrial gas leakage site situated at Bhopal, Madhya Pradesh, India	132
7.2	Death rates from outdoor air pollution, 2017	136
7.3	The Mekong river basin	140
7.4	Women collecting water near Mzuzu, Malawi	142
7.5	Water and food insecurity intersect to cause health problems	143
7.6	A tubewell in Bangladesh where water is contaminated with arsenic	147
8.1	Stagnant water as breeding ground for *Aedes* mosquitoes	158
8.2	Spatial variation in malaria linked to two parasite species in Brazil, 2018	160
8.3	Ebola virus disease cumulative incidence, 20 September 2014	161
8.4	Caged birds for sale in a wet market in Shanghai, China	164
8.5	Location of patients with confirmed Covid-19 in China, 19 February 2020	167
8.6	Global inequality in number of Covid-19 vaccine doses administered by April 2021	171
9.1	Relative risk of heat-related mortality per 10°C increase above the 95th percentile observed daily temperature	178
9.2	Possible shift of infections transmitted by *Aedes aegypti* mosquitoes from 2020 to 2050 to 2080	179
9.3	Pathways of climate change, food security and maternal/infant health reported by mothers in rural Uganda	183
9.4	Conceptual model of climate, conflict and migration	185

LIST OF FIGURES, TABLES AND BOXES

Tables

2.1	Sustainable Development Goals	23
2.2	Life expectancy in selected sub-Saharan countries, 2019	26
2.3	Health and social indicators in Brazilian regions, 2013	31
2.4	States in India with health indicators	33
2.5	Child mortality by province in Mozambique, 2003	39
3.1	Percentage of male adults smoking, from SDG Target 3.a, 2007 & 2018	47
3.2	Total alcohol consumption by persons over 15 years (in litres, per capita), from SDG Target 3.5, 2000 & 2018	48
3.3	Age-standardized years of life lost per 100,000 in China, 2017	49
3.4	Age-adjusted percentages of Native American adults in fair or poor health, by area type, 2014–18	53
3.5	Disability scores in contrasting census areas in Rio de Janeiro, 2006	57
5.1	Global forced displacement, 2020	95
6.1	Blood lead levels (in μ/dL) in a town handling e-waste (Giuyu) and a comparison town (Chendian), China	124
8.1	The major neglected tropical diseases	154
8.2	Estimated excess mortality rate (per 100,000), and reported Covid-19 deaths, by country/state, 2020/21	168
8.3	Countries with fewer than 5 per cent of the population fully vaccinated, April 2022	170
10.1	Cross-cutting themes in health geography and global health	189

Boxes

1.1	Structural violence	3
1.2	What's in a name?	5
1.3	The political ecology of health among 'Namgis First Nations in British Columbia	6
1.4	Assemblages and the post-human turn in health geography	9

1.5	Universal truths and situated knowledge	15
2.1	Political and commercial determinants of health	21
2.2	The Uighur population in Xinjiang province	37
3.1	New markets for smokers	48
3.2	The Gender Inequality Index	50
3.3	Historical trauma	54
3.4	Antidepressants in the Global South	61
3.5	Child mental health in Gaza	62
3.6	The inverse care law applies in the Global South too	67
4.1	The Alma Ata conference (1978)	75
4.2	The Washington Consensus	77
4.3	Getting TRIPed up	79
4.4	The 2014 Ebola outbreak and the role of MSF	83
4.5	The Bill & Melinda Gates Foundation (BMGF)	85
4.6	Global health surveillance	89
4.7	The Global Health Security Index	90
5.1	The Internal Displacement Index	96
5.2	Fleeing Sudan	98
5.3	Attempting to cross the English Channel	98
5.4	Rohingya refugees in Bangladesh	101
5.5	One woman's search for safety: Zarlasht Halaimzai	104
5.6	Ontological security and the emotional health of refugees	106
5.7	Every child protected?	110
6.1	Global Environmental Justice	115
6.2	Pollution havens	121
6.3	From structural violence to slow violence	125
7.1	A story from Bhopal	133
7.2	Transboundary issues in the Mekong river basin	141
8.1	A syndemic perspective	154
8.2	Zika virus	157

8.3	Antimicrobial resistance	165
9.1	The coloniality of climate change	176
9.2	Climate change, population displacement and political conflict: a cautionary tale	184
9.3	Adapting to climate change: a critical perspective	186
10.1	A happy planet?	191
10.2	Health and human rights	193

1
INTRODUCTION

Clearly, many disciplines, such as the social and behavioural sciences, law, economics, history, engineering, biomedical and environmental sciences, and public policy, can make great contributions to global health.

Koplan *et al.* (2009: 1994)

Geographies of health

Human geography is the study of human activity in relation to the places and environments that people inhabit, and health geography seeks to describe and explain how health (and wellbeing, and illness and disease, and the determinants of such) varies from place to place (Gatrell & Elliott 2015; Brown *et al.* 2018). Places can encompass diverse settings, including the home, workplace, neighbourhood, region and country. As a result, the subject matter spans a considerable range, such as the experiences of those living at home with chronic illness, how access to resources affects the wellbeing of those living in different neighbourhoods, and regional inequalities in morbidity and mortality. Most importantly, as I hope to show in this book, relations between (and within) different nation states; the flows of people, materials and infections, along with the historical, political and economic linkages between them, are – or should be – a focus of health geographers' attention. Geography, as a distinct discipline, should be added to Koplan *et al.*'s disciplinary mix.

Approaches to the study of health geography are diverse, as revealed in Gatrell and Elliott (2015: ch. 2), the breadth of essays in the collection edited by Crooks, Andrews and Pearce (2018) and the critical introduction to the subject by Brown *et al.* (2018). Until the 1990s geographical enquiry that focused on illness and disease was labelled "medical geography" (Mayer 2010 has an excellent overview). Although some of that research looked at the distribution of morbidity (illness) and mortality in advanced economies, other research dealt with tropical diseases such as malaria. That tradition was called "disease ecology" (Learmonth 1988) – discussed more fully below – and it explored the ways in which the incidence of disease was associated with environmental conditions. Paralleling this was a rich tradition of mapping and analysing the distribution of mortality and morbidity (as well as the spread or diffusion of disease) that led to sophisticated visualizations and spatial analysis, often linked to the use of geographical information systems.

It took a seminal paper by Robin Kearns (1993) to suggest that more attention needed to be given to health, as opposed to disease. Out of this grew a new body of work, more concerned with health in specific social and cultural settings and less with the mapping and analysis of spatial variation in disease. As a result, some health geographers seek to understand how individuals experience (ill) health in different places, drawing on qualitative methods such as interviews, diaries and focus groups. For example, Milligan (2018: 230) considers research on the home a "space imbued with multiple meanings linked to identity, safety and security, power and control, emotion, nurture and historical memory". However, such research has tended to relate to homes in the Global North rather than to the safety and security in the dwelling places of those living in countries of the Global South.

A further broad approach to health geography situates health in the wider context of social and economic conditions – a structuralist or political economy perspective. This can draw on a variety of methods, both quantitative and qualitative, but it frequently demands listening carefully to the voices of those usually drowned out by the views of experts and policy-makers. The experiences, beliefs and rights of those living under the constraints of health policies imposed on them provide a body of knowledge that deserves attention. For Bambra, Smith and Pearce (2019: 38), political economy means that population health is shaped by the "social, political and economic structures and relations that may be, and often are, outside the control of the individuals they affect". They note that such insights have yet to be fully developed and applied within the literature on health and place.

Although not the exclusive preserve of anthropology, it is that discipline that has majored in a political economy of health and done most to illuminate global health from the perspective of ordinary people. Excellent examples include the rich ethnographic research undertaken on people living with HIV in Haiti (Farmer 2005) and Ebola in west Africa (Farmer 2020). Chapters in the collection edited by Biehl and Petryna (2013) also illustrate the importance of ethnography in revealing the voices and stories of people living, for example, with malaria and AIDS. Biehl and Petryna (2013: 3) argue that "looking closely at life stories and at the ups and downs of individuals and communities as they grapple with inequality, struggle to access technology, and confront novel state-market formations, we begin to apprehend larger systems". As this (and Farmer's work) suggests, ethnographic approaches are essential in a political economic approach. A key concept in Farmer's work is that of structural violence (Box 1.1).

BOX 1.1 STRUCTURAL VIOLENCE

By "structural violence" is meant not physical violence (although, as we see later, that is real enough for many in the Global South) but the violence perpetrated by power relations that dictate who lacks access to the basic resources necessary to sustain life. Suffering is structured by economic and political forces (racism, sexism, political instability and macro-economic policies) to shape and constrain human agency and health. The axes of gender, race and poverty intersect to dictate the health of individuals in specific geographical settings.

One of Farmer's several books, *Pathologies of Power* (2005), illuminates in particular the suffering endured by the poor living in Haiti, a country he describes graphically as "a sort of living laboratory for the study of affliction" (2005: 30): see Figure 1.1. As he notes, we are moved by the suffering of those close to us (think of the impact on health and livelihoods of those affected by severe flooding in mainland Europe, or the displacement of people from Ukraine, or those deeply affected by the hurricanes making landfall in Louisiana, for example) but less moved by the suffering of those who are remote, whether in geographical or cultural terms. To explain such suffering, Farmer exhorts us to connect individual biographies with political economy.

Büyüm *et al.* (2020) write that the Covid-19 pandemic (see Chapter 8) has highlighted clearly how structural violence operates both within and between countries. Structural violence gives precedence to some social groups while disadvantaging and harming others. Herrick (2017a) has drawn on the concept of structural violence in her study of alcohol consumption in South Africa. She sees alcohol consumption in poor urban communities as a form of coping with the structural violence caused by racial segregation, inequality and unemployment.

For a detailed examination, and critique, of structural violence, see Herrick and Bell (2022).

The recent textbook on health geography (Brown *et al.* 2018) is subtitled *A Critical Introduction*. By this, the authors mean a key concern with issues of social justice and power relationships. Crucially, and suggesting sympathy with a political economy perspective on the subject, they suggest that a critical approach means paying close attention to those who are marginalized, whether in the Global North or Global South (Box 1.2).

As I have indicated, there has been a discernible shift away from medical geography towards "health geography" – or, more accurately, *geographies*, as the titles of two textbooks suggest (Gatrell & Elliott 2015; Brown *et al.* 2018). But, to date, there has been a relative neglect in that subdiscipline of the bigger picture of global health. This is evidenced by scanning the literature on key current topics. For example, literature on therapeutic landscapes, geographies of ageing and of children, mental health,

Figure 1.1 Haiti as a "living laboratory for the study of affliction"
Source: iStock.

home environments, walkability, green/blue spaces, and rural health – all considered in a recent edited volume (Crooks, Andrews & Pearce 2018) – have relatively little to say about such issues as they pertain to the Global South. The same may be said about the current interest in wellbeing. As Severson and Collins (2018: 128) put it, "Implicit in [that] research is a recognition that well-being – in terms of overall contentment and quality-of-life beyond satisfaction of basic material needs – is something the relatively privileged can afford to work on."

Despite the move away from medical to health geographies, a focus on disease (particularly infectious disease) rather than health has always been a concern of some geographers. Good examples are the body of work undertaken in Britain by Cliff, Haggett and Smallman-Raynor (see, for example, Cliff *et al.* 2009) and ongoing work by those who focus their attention on disease ecology.

Disease ecology suggests that, to understand the geographical distribution of a disease, we need to link together habitat, population and human behaviour. "Habitat" includes both the physical, biological and built environments. In a riposte to Kearns' seminal paper (1993) referred to above, Mayer and Meade (1994: 103) say that disease ecology "considers the numerous social, economic, behavioural, cultural, environmental, biological factors which create disease in specific places at specific times". All-embracing as this seems to be, in a later paper Mayer (1996) stresses the political dimension and introduces the political ecology of disease as "one new focus" (as his title puts it) for medical geography.

BOX 1.2 WHAT'S IN A NAME?

Throughout this book I use "Global North" and "Global South" as simple categories, but, as Khan and colleagues (2022) have noted, there is a wide variety of terms – often rooted in colonial thinking ("First World" and "Third World", for example) – that have been used to classify major regions.

The distinction between Global North and Global South has its origins in the Brandt Report of 1980 (Figure 1.2), but, "given the complex ways in which globalizing capitalism is reterritorializing human development, poverty, and inequality today, the Brandt Line is becoming less and less fruitful as a heuristic" (Boyle 2021: 195). Boyle (2021) goes on to suggest (197) that "the First World can now be found in the Third and the Third World in the First". Some areas and communities in the Global North reveal patterns of inequity and ill health that are not so different from those in the Global South.

Lencucha and Neupane (2022) argue that classifications such as "low- and middle-income countries" (LMICs) risk perpetuating hierarchies, constructing one group as subservient to another that extracts wealth and resources from it. Acknowledging the dangers of simple classifications, I use the terms "Global North" and "Global South" (and, given their widespread use in the research literature, "low- and middle-income countries") as convenient labels, but focus attention on people and places that are peripheral in terms of their relative location in social, economic and political spaces.

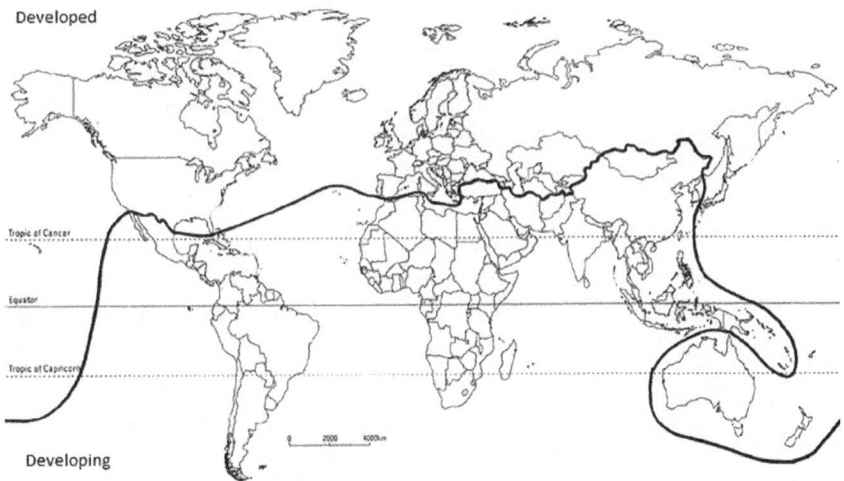

Figure 1.2 The "Brandt line", separating the Global North ("developed") from the Global South ("developing")
Source: Courtesy of Mark Boyle, Maynooth University, Ireland.

Political ecology and health

In understanding the geography of disease and ill health, political ecology conjoins political economy with classical disease ecology, to reveal how economic and political processes and structures interact with environmental processes and constraints to produce ill health. All these processes, structures, determinants and distributions (of ill health) are played out unequally in different parts of the globe. Mayer (1996) gives an example of a political ecology approach, referring to classic research on malaria in the first half of the twentieth century in British Malaya, where large-scale clearance of the jungle had enabled large corporations to build rubber plantations. The removal of forest cover, associated with stagnant pools of water, allowed *Anopheles* mosquitoes (those carrying the parasite that causes the disease) to breed, exposing workers to disease risk.

In a seminal paper, Chantelle Richmond and her colleagues (2005) adopt a political ecology framework to understand aquaculture in British Columbia, and, as this links a number of themes in the present chapter and the next, it is worth considering more fully (Box 1.3).

BOX 1.3 THE POLITICAL ECOLOGY OF HEALTH AMONG 'NAMGIS FIRST NATIONS IN BRITISH COLUMBIA

Aquaculture – the cultivation and harvesting of fish – is an important contributor to the economy of Canada; about 7 per cent of farmed salmon worldwide is provided from the country, making it the fourth largest producer across the globe. Richmond and her colleagues (2005) note that the industry, centred in parts of coastal British Columbia, has become globalized, with resulting impacts on First Nation communities (specifically, 'Namgis First Nation), for which fishing has been a way of life for hundreds of years. For these communities, and other Indigenous groups, health and wellbeing are embedded in the local environments they inhabit. The threats of commercialized fishing include both environmental and health risks, and Richmond's research seeks to understand the communities' own perceptions of these risks. The legacy of colonialism in Canada (*British Columbia*) means that First Nations' claims to environmental resources have long been dismissed.

Detailed qualitative research uncovered links between resource use, economic opportunities, lack of autonomy and health outcomes. As one respondent, Robbie, says:

> There are always concerns about the health of our people and I think the health of the environment goes a big way towards the health of our people too. I think if your forests and the oceans die, I think the people are going to die along with

Figure 1.3 Musgamagw Tsawataineuk protests at a salmon fish farm
Source: Kwakwaka'wakw: www.firstnations.de/fisheries/kwakwakawakw.htm (courtesy of Karen Wonders, University of Göttingen, Germany).

it. There have got to be measures and steps taken to ensure that we have a healthy land and healthy people.

Marianne comments: "We are finding out more about effects from the chemicals to make [salmon] flesh pink ... They might have side effects on humans ... the antibiotics they are given, might have side effects on humans" (quotes from Richmond et al. 2005).

The political ecology framework adopted in this research helps to link the local environmental context, and population health, to broader national and international structures. As the authors' research demonstrates (Richmond et al. 2005: 352), "The theoretical constructs and methodological framework of 'third world ' political ecology are equally applicable to peripheral regions within advanced capitalist economies, for instance marginalized populations of the First World such as First Nations communities." Their research brings to the fore how political economy connects with environmental extraction to lay bare the health inequities faced by an Indigenous population (Figure 1.3). Such inequities are considered in detail in Chapter 2 and elsewhere in the book.

See Richmond et al. (2005) and Richmond and Big-Canoe (2018) for additional detail.

What happens in particular places cannot be divorced from the wider economic and political context; all scales are "mutually enmeshed" (Mayer 1996: 447). Presciently (prior to Covid-19), Mayer's important review discusses the value of a political ecology approach for understanding emerging and re-emerging infections – a theme to which I return in detail in Chapter 8. For example, the emergence of Lyme disease in the United States has been widely linked to the encroachment of suburban housing on previously forested areas, bringing people into contact with animals harbouring the ticks that carry the infectious agent. Mayer also reminds us of the health consequences of constructing the Aswan Dam, a project that was part of Egypt's drive to post-war industrialization, whereby the accompanying lake behind the dam provided a breeding ground for the snails transmitting schistosomiasis to local populations (see Lerer and Scudder 1999 for a thorough review of the health impacts of large dams, and further discussion in Chapter 7).

Mayer has been criticized for seeming to give more attention to disease ecology than to political economy. As Jackson and Neely (2015: 50) put it: "In political ecology, the insights of critical geography, especially Marxism, are constitutive of the problems scholars seek to research and the methods they choose; quite simply, without critical geography, there would be no political ecology." The authors make explicit the links to Marxism and feminism, in which inequalities of capital and gender come into play, revealing how uneven global political and economic processes manifest themselves among those in specific local social and cultural contexts.

Nichols and Del Casino (2021) develop these ideas further, asking what the role is that affect and emotion play in political ecology. They reflect the introduction into health geography of poststructuralist approaches, such as non-representational theory (NRT: Andrews 2018). Here, "affect and energy both provide connective imaginations that indicate how different elements, objects, bodies and surfaces might be brought into inter-relation toward the constitution of health or ill-health" (Lea 2018: 148). Lea suggests that NRT helps "the reconfiguration of health in terms of pleasure and seduction, rather than illness and disease", although she adds the rider "*not, by any means, for everyone in society*" (emphasis added). If, as she argues, NRT moves the focus of health geographers away from illness and suffering, it raises the question as to how NRT might illuminate the lives and circumstances of the vast majority of humanity. It remains to be seen whether attention given to affect and emotion has anything substantial to contribute to the study of global health when illness and suffering mostly trump a concern with "pleasure and seduction".

This is not to downplay the importance of understanding the human body in political ecologies of health. Nichols and Del Casino (2021) show very clearly how this is important in our understanding of diabetes in India. For these authors, the body is an "historical socio-biochemical assemblage and also an affective/nonrepresentable happening" (787). They are not the first to introduce "assemblage" as a key concept in critical health geography (Box 1.4).

BOX 1.4 ASSEMBLAGES AND THE POST-HUMAN TURN IN HEALTH GEOGRAPHY

The last 20 years have witnessed significant interest among geographer in "more-than-human" geographies, a poststructuralist perspective that does away with binary oppositions of agency/structure, inside/outside, natural/social and instead asserts that human and non-human factors or forces come together in complex networks or "assemblages". The latter word is most associated with philosopher Gilles Deleuze, for whom "the term assemblage describes the 'co-functioning' of heterogeneous parts into some form of provisional, open, whole. An assemblage is a 'harlequin's jacket or patchwork' of different bodies that can never be reduced to a series of constituent parts nor identified as an organic whole" (Anderson *et al.* 2012: 177). Put more simply, non-human entities – not only animals, but even technologies – are regarded as having "agency", alongside the human beings with which they network. Such a view matches that taken in science and technology studies (STS), championed by Bruno Latour and others, where things cannot be categorized as either natural or socio-cultural but are actors within networks. Some anthropologists argue that the distinction between human and non-human beings, foundational to Western thought, is not one made outside the West (Rock, Degeling & Blue 2014). For others, a more-than-human or "post-human" perspective "brings into question the very idea of health as the normal condition" (Duff 2018: 140). It also critiques research on health inequalities, in which "social and structural factors such as power, class, income inequality and gender are *said to have* a causal effect on the health of populations" (Duff 2018: 140, emphasis added).

How does this relate to political ecologies of health? For Neely and Nading (2017: 58), it means that "human bodies are not simply 'exposed' to pathogens or hazards or risks in particular locales; rather, they 'incorporate' the nonhuman activity into their own metabolic and immunological processes". Or, as Duff (2018: 140) puts it, "it is not a space, a pathogen, a practice or an attitude that makes an individual subject ill; rather, illness emerges within an assemblage of human and nonhuman forces". In other words, it is the assemblage that gets ill.

In health geography, the concept of assemblage is, like non-representational theory, just one aspect of poststructuralist views of the world. In health geography it requires taking the body seriously as a "site" (even "place") of investigation and noting the position of the researcher in relation to their subject matter (Evans 2018). Further, poststructuralism is also sceptical about the universality of knowledge (Evans 2018), a point that is elaborated below. Yet, as Evans points out, poststructuralism can downplay the role of the intentional human actor, and he questions the relevance of poststructuralist accounts to the policy-maker.

For further consideration of poststructuralist approaches and more-than-human health geographies, useful starting points are Duff (2018) and Evans (2018). See also the paper by Rock, Degeling and Blue (2014).

Emotion and affect certainly merit attention in political ecologies of health. A good example of this is important research by Farhana Sultana (2011, 2012) on arsenic contamination of water in Bangladesh (considered further in Chapter 7).

An acknowledgement of the importance of emotions and affect should not deflect attention from the role of political economy in political ecology, and nor should contemporary engagement with assemblages mask the fact that networks of human and non-human, as well as place, have always been part of classical disease ecology. For example, Neely and Nading (2017) discuss the mosquito-borne Zika virus and the accompanying incidence of microcephaly (when a baby's head is smaller than expected) in Brazil, where both place and inequality came together. The incidence of microcephaly was highest in one of Brazil's most poor, and socially segregated, areas, and preceded the emergence of Zika virus. Later chapters discuss Zika virus more fully.

Globalization

Although there are many possible definitions of globalization, the following is useful and has obvious relevance for global health: "Globalization is a complex set of social, economic, technological, political, and cultural developments that have created new connections and interdependencies between people, places, and environments" (Liverani 2015: 155). The present book is an attempt to examine these interdependences and connections as they relate to human health, although, as argued already, and examined fully in Chapters 2 and 3, an examination of global health demands attention to health inequalities.

Interdependences and connections across the globe are nothing new. There have, for centuries, been trading links and colonizations, with consequences for health. Most obviously, the word "quarantine" (isolation for 40 days) derives from attempts to control the spread of plague by traders and travellers in fourteenth-century Venice, while infections such as smallpox were deliberately introduced by those colonizing the New World in the seventeenth century (Gatrell & Elliott 2015: 44–5). But the emergence of new transport technologies in the nineteenth century, and subsequent air transport in the twentieth century, has led to "time–space compression" (also called space–time convergence), a process by which places "converge" or are brought closer together in time-space (Figure 1.4) (Gatrell 2011). These technologies have led to a speeding up of exchange and connection, whether of finance, material goods, or people and the diseases they may carry with them.

Globalization ensures that what happens in specific, local places may be shaped by events hundreds of miles away (Urry 2003). However, in a study of global health it cannot only be events occurring many miles away that shape our study. As we see later in discussing infectious disease, events in one place can be affected dramatically by what is happening just a few miles away, such as in adjacent and ill-defined borders. The global and the local are, therefore, interdependent. For example, locally high rates of tobacco consumption depend on the actions of global corporations, while local outbreaks

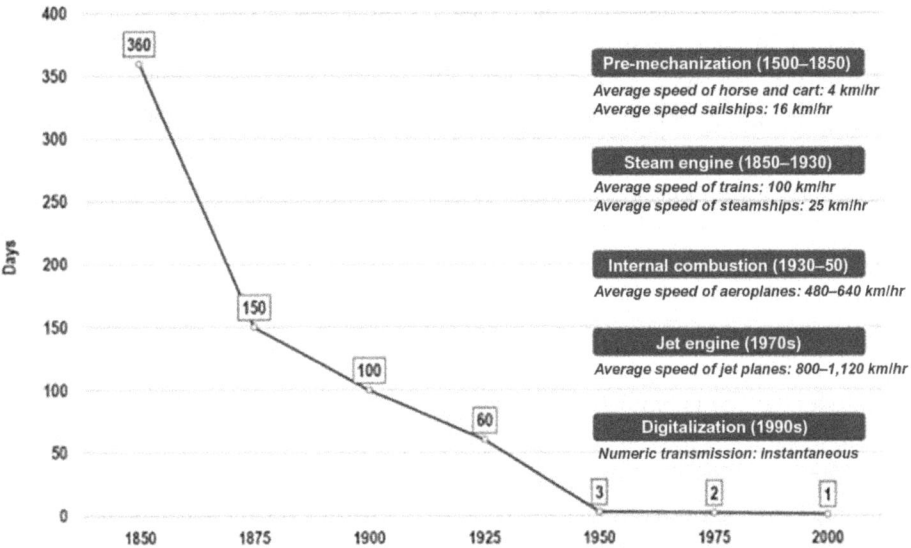

Figure 1.4 Space–time convergence
Source: Global space/time convergence: days required to circumnavigate the globe: https://transportgeography.org/contents/chapter1/transportation-and-space/world-circumnavigation-days (courtesy of Jean-Paul Rodrigue, Hofstra University, United States).

of disease are linked to wider-scale changes in climate, and spatially specific poor health outcomes are not independent of austerity measures imposed by the policies of major international bodies such as the World Bank.

Clearly, in globalization it is movement that is key, whether of goods (or "bads"), information, people – or infections. What matters in the global economy is networks – of businesses, corporations, enterprises, countries and regions. But not everyone has access to these networks; many people do not reap any benefits from globalization. Some people and places are indeed bound together tightly into flows and networks, while others get bypassed and overlooked. There is an unevenness to globalization: inequality or inequity is cemented in. As Navarro (1999: 220) puts it: "[G]lobalization is not neutral. There is nothing intrinsically good or bad in the flow of capital, labour, and knowledge around the world; its goodness or badness depends on who governs the flow, which determines who benefits from it." More stringent critiques of globalization (such as Veseth 2005) refer to it as "globaloney", arguing that distance and borders still matter; Veseth prefers the term "semi-globalization". Sassen (1996: xvi) acknowledges the "denationalisation of economic space" (the liberalization of trade) while at the same time referring to the "renationalizing of politics" (the policing of national borders to control population movement).

The sociologist Ulrich Beck links globalization to what he terms "second modernity" (Beck & Lau 2005), in which modern industrial and technological change characterizes a "risk society" – broadened out by Beck (1998) as a "world risk society" (or global risk society). In essence, he argues that technological and scientific progress is accompanied by increasing levels of risk. Such risks include, in Beck's writing, climate change, pollution, nuclear energy and the spread of infectious disease. As we see later in the book, some infectious diseases have threatened at one stage to pose global risks, while in other cases (Covid-19, of course) the threat has become devastating worldwide. Later chapters show how the movement of materials, air and water pollution, and climate change, also have dire consequences for the health and wellbeing of people across the globe.

Global health and global health geographies

Kickbusch (2006: 561) has described global health as the study of "those health issues that transcend national boundaries and governments and call for actions on the global forces that determine the health of people". This suggests a particular focus on disease, ill health and their determinants that are not confined to single nation states. What happens at, and across, international borders matters, as does the role played by non-state actors and institutions, non-governmental organizations (NGOs) and private philanthropists (see Chapter 4). In a key introductory text, Farmer and his colleagues (2013) want to include the study of health disparities or inequities, both between and within countries. Koplan and colleagues (2009) agree. The study of global health is not to be restricted to health-related issues that *literally* transcend international borders. Inequality and social justice are, or should be, key concerns, particularly in the Global South, as well as transnational determinants. This is echoed by Herrick (2016: 674), who suggests that global health coalesces around "the will to secure health for all in the most efficient way possible, redress fundamental inequities, deliver value for money, and tackle the mounting economic burden and security threats posed by disease and suffering".

Having said that, global health covers not only those countries formerly considered to be "developing". As Meyers and Hunt (2014: 1922) put it: "Access to clean, affordable water is an issue that joins the challenges of providing a sanitary infrastructure for poor, rich, and the middle class in Detroit, Delhi, Lagos, and Johannesburg alike." In measuring rates of disease and illness, and income inequality, cities such as Detroit are not so dissimilar from places in the Global South. Herrick and Reubi (2007b: xv), in the introduction to their edited collection, make the same point about healthcare, reminding us that good healthcare exists in many parts of the Global South, just as there are places of seriously poor access in parts of the Global North. We need to avoid casting all those living in poorer countries as victims: countries, regions and communities are invariably heterogeneous rather than uniformly unhealthy.

Recent outbreaks of serious infectious disease, such as Ebola and Covid-19, risk constructing global health largely in terms of communicable disease, not least because the infections themselves are invisible, even if the consequences are far from it. Herrick and

Reubi (2017b: xviii) remind us that a key task for global health geography is to attend to non-communicable disease (NCD) and its determinants; this is explored further in Chapter 3. In addition, contributors to their book (Herrick & Reubi 2017a) flag the importance of movement, whether this be of body parts (sometimes called "transplant tourism") or of health professionals from one country to another. Elsewhere, contributors to the Herrick and Reubi collection adopt the more-than-human perspectives referred to above. For example, in her chapter on malaria, Beisel (2017) calls for "multispecies understandings", telling us that "people who regularly live with malaria have developed their own ways to negotiate living well and staying free of malaria parasites, or treating infections quickly and effectively" (130). Echoing the concept of assemblage (Box 1.3 above), she speaks of a relational web of actors, human and otherwise (patients, clinics, bed nets, drugs and mosquitoes), that need to be understood. Hinchliffe (2017) makes a similar point, although I maintain my earlier observation that disease ecology – emerging long before poststructuralist approaches became popular in human geography – was entirely focused on such entanglements.

Relational webs and networks are prioritized by other geographers writing on global health. Brown and Taylor (2018: 14) suggest that, if we are to address global health issues, particularly those relating to inequality, we need to bring together "ministries of health, physicians, patients and super-rich donors in a complex web of multinational institutions, bilateral partnerships, and advocacy networks". Correspondingly, like Beisel and Hinchliffe, the authors want to bring to bear on global health the more-than-human turn, stressing the need for coexistence with "the much-maligned bugs, viruses, bacteria and parasites that live within and beyond the human body" (Brown & Taylor 2018: 17). The "One World, One Health" initiative (Calistri *et al.* 2013; see also Andrews 2021) is an example of an initiative to promote study of the interdependence of all ecosystem components, whether locally, nationally or globally, in order to understand human health.

If these are some of the themes with which health geographers are to engage, how should they go about addressing them? What principles should be adopted for doing global health geography? This is considered by Plamondon and Bisung (2019), who propose six principles for undertaking research on global health. These are: authentic partnering; inclusion; shared benefits; commitment to the future; responsiveness to causes of inequities; and humility. In brief, research partnerships need to be based on an equal footing and to include the marginalized groups being studied. The benefits of such research need to be shared, such that the contributions of those groups are recognized and valued and research skills are transferred to those who can use them in future projects. Addressing inequity is foundational. The Canadian Institutes of Health Research discuss this comprehensively (CIHR 2021).

Adopting such principles might go some way towards addressing the concerns expressed by Abimbola and Pai (2020), who assert that the discipline of global health "holds within itself a deep contradiction – global health was birthed in supremacy, but its mission is to reduce or eliminate inequities globally". They want it to become anti-supremacist and anti-racist. In essence, they want it to "decolonize".

Decolonizing geography and global health

Imperial projects of tropical medicine helped white colonialists settle in, and exploit, what is referred to as the Global South. A good example comes from Freetown in Sierra Leone, where the British Colonial Office sought, in the later nineteenth century, to protect government officers from malaria by segregating the white population from the local community by constructing hill stations away from the town centre, adjudged sufficiently distant to avoid mosquitoes breeding in stagnant pools that had been identified as sites of risk (Frenkel & Western 1988). This strategy of residential segregation embodied the sense in which white Europeans regarded themselves as far superior to those they sought to dominate. The emerging science of tropical medicine was used to protect one population rather than help improve the living conditions of the "other". Boundaries were drawn, and maintained.

More generally, the ways in which the history of geography as a discipline has been wrapped up in an imperial project are well known (see, for example, Bell 1995), but only recently have geographers become attuned to decolonization (Boyle 2021: 409–14). For some writers, the task is a challenging one. "Putting it baldly, geography's history is of a terrible and problematic opening out *of* the world to colonial and exploitative forces, and the continued whiteness of its post-colonising heartland displays little *practical* contemporary openness to difference and diversity in its knowledge production processes" (Noxolo 2017: 317). Ferretti (2020: 1167) is more optimistic, suggesting that the discipline is becoming more inclusive of those working outside Anglo-America, and the perspectives they bring to bear. However, as he recognizes, such scholarship remains largely in the hands of those speaking English (Craggs 2019).

Radcliffe (2017) agrees with Noxolo, suggesting that, despite the emergence of independent national states after divesting themselves of colonial rule, forms of knowledge have remained "deeply rooted in post-Enlightenment Euro-American claims to be able to pronounce *universal truths* and to theorise the world" (Radcliffe 2017: 329, emphasis added; see Box 1.5). In contrast, the decolonial turn draws on perspectives from outside Europe and North America, calling for the voices to be heard of Indigenous people (Richmond & Big-Canoe 2018; Eni *et al.* 2021) and those marginalized in the Global South.

Some writers have taken it upon themselves to actively address the decolonization agenda in global health; see, for example, Herrick, Okpako and Millington (2021) and the Decolonising Geography Educators Group (www.decolonisegeography.com). Oti and Ncayiyana (2021) have established the Global Health Decolonisation Movement in Africa to give a stronger voice to Africans working in this field. They acknowledge the challenge, suggesting they risk criticism from "those who do not even subscribe to the idea of or need for decolonization in the first place. Just like the *#BlackLivesMatter* movement was rebuffed cynically with *#AllLivesMatter*, we anticipate that the

BOX 1.5 UNIVERSAL TRUTHS AND SITUATED KNOWLEDGE

Among other writers, feminist scholar Donna Haraway has criticized claims of scientific objectivity, suggesting that science favours *"supposedly objective* observational technologies which claim to produce a truthful account of the world" (Holloway 2004: 168, emphasis added) over subjective human observation. Importantly, from a decolonial perspective, she asserts that what she calls "situated knowledge" should include that produced by subjugated groups, whether these be women, those with disabilities or those of colour. Knowledge and engagement with nature, technologies, and health are multiple and varied, and Haraway demands that this fact be acknowledged.

How does this relate to global public health? In a recent commentary, Affun-Adegbulu and Adegbulu (2020) write that it "presupposes a notion of a universal human subject", and tends to reject work that does not conform to that notion. Echoing Haraway's situated knowledge, they suggest that objectivity is impossible and that we should engage with multiple perspectives on the world. The same points are raised by those working in the field of international relations. For example, Querejazu (2016: 2) writes: "[T]he ultimate 'truth' of one-world, one reality and one universe is a myth." Instead, drawing on her work with Andean peoples, she counsels that we should consider a "pluriverse", the many kinds of knowing reality (Escobar 2020). "One world reality", as Querejazu puts it, has been imposed on the "other". Indigenous knowledge, such as that of the Andean, puts humans in the role of environmental caretakers, not dominant beings. The human world, the natural world and the spiritual world all coexist.

This is echoed in the biting critique by Abimbola and Pai (2020): "Supremacy is seen in persisting disregard for local and Indigenous knowledge, pretence of knowledge, refusal to learn from places and people too often deemed 'inferior', and failure to see that there are many ways of being and doing." Dominant cultures need to be decentred in order to allow marginalized actors and communities to engage in knowledge construction. Adoption of the principles for doing global health research might go some way towards alleviating these concerns.

#DecoloniseGlobalHealth movement will continue to be met with resistance" (Oti & Ncayiyana 2021: 2).

Stark criticisms such as these have a strong bearing on the efforts of governments, NGOs and wealthy individuals to address global ill health. Some writers (such as Chaudhuri *et al.* 2021) see such efforts simply as another means of exercising Western power and domination. They argue that the global health industry continues to smack of colonialism. I return to this issue in Chapter 4.

Concluding remarks

The brief overview here of approaches to *doing* health geography suggests that many of these can help inform our understanding of global health, even though much of the contemporary literature has yet to concern itself with the "global". To understand what happens where, and why, particularly in the Global South, the main focus in this book, we need a wide range of tools. We need ethnographies to hear the voices of those who are often silenced, but, equally, there can be a place for intelligence provided by survey data, mapping and the analysis provided by geographical information systems/science (GISc) – although this has to be in partnership with those who can benefit from such technologies. The tools we use depend on the questions we are posing.

We have seen that the subdiscipline of medical geography has broadened out to health geography but that the focus on disease ecology by some medical geographers has itself expanded to a rich focus on the political ecology of disease. Over the last ten years health geographers have drawn on poststructuralist perspectives to bring to the fore a concern with emotion and affect.

The chapter has introduced ideas about globalization and identified some of the fundamental concerns of global health. The chapter ends by considering the decolonization of global health and the importance of acknowledging non-Western perspectives on health. Global health and its geographical connections cannot but help be a political enterprise that addresses the concerns of marginalized groups. The Marxist geographer David Harvey wants to "reject a de-politicized geography that eviscerates any mention of class and class struggle or of capital accumulation at a time of intense restoration and reconstitution of class power everywhere around the globe" (Harvey 2006).

The remainder of the book builds on some of the ideas in this chapter. I begin by looking at health inequalities, then at how global health is governed, before looking in depth at aspects of the movement of both peoples and materials, and the health consequences that derive from such mobilities. I then turn to associations between air and water quality and human health before considering infectious disease (the incidence, spread, determinants and consequences of such), and diseases that are "neglected". Last, the health impacts of climate change are considered.

In covering all these topics, I want to stress how these different issues play out in particular geographical settings, with an emphasis (albeit not an exclusive one) on countries in the Global South and where they cut across national borders.

Further reading

The recent introduction to human geography by Boyle (2021) is a comprehensive and engaging introduction to the subject, touching on many of the themes in the present book, including inequality, population movement and infectious disease. I cannot recommend it too highly.

For introductions to the variety of topics and approaches to health geography, see the excellent collection of 52 essays in the handbook edited by Crooks, Andrews and Pearce (2018). Gatrell and Elliott (2015) also introduce the subject, albeit without the breadth given in an edited collection of 52 chapters. More recently, the set of chapters in Brown *et al.* (2018) provides the perfect background to what follows, and I suggest strongly that, if new to the geography of health, you start there. Luginaah and Bezner-Kerr (2015) have edited a valuable collection of essays on health and development.

On global health and globalization and health, you should see the collection of essays in Kawachi and Wamala (2006a). There are examples of geographical research on global health in Herrick and Reubi (2017a), but, for an excellent discussion of global health from a geographical perspective, see Neely and Nading (2017). Many relevant research papers on global health appear regularly in the journals *BMJ Global Health*, *The Lancet Global Health* and the *Journal of Global Health*, as well as more journals of direct relevance to health geographers, such as *Health & Place* and *Social Science & Medicine*.

For an overview of the political ecology of health and disease, see Kalipeni, Ghosh and Oppong (2017), but also the *International Handbook of Political Ecology*, edited by Bryant (2015). The collection edited by Rocheleau, Thomas-Slayter and Wangari (1996) is an important resource on feminist political ecology.

The recent book by Amzat and Razum (2022) covers many of the topics discussed in the following chapters, and its chapters on globalization are essential reading.

2
UNEQUAL HEALTH I: DETERMINANTS AND REGIONAL EXAMPLES

> Great and growing global inequity, the burden of poverty both absolute and relative, millions of preventable deaths every year – these unsettling features of today's world lead many students toward global development and health work because it seems like the only decent thing to do. Suri *et al.* (2013: 245)

Brown and Taylor (2018) suggest that the study of health inequalities needs to be at the centre of geographical research on global health. It is therefore appropriate to begin an exploration of global health by describing, and explaining, the health inequality (more accurately, health *inequity*) that exists between and within different countries. "Health inequality" is invariably the term in common use, and so I follow that trend, although "health inequity" conveys more clearly that differences in health may be preventable – avoidable and unjust. The issue of who or what are responsible for avoidance – human agency, or wider structures of economy and politics – is still debated, although a critical geographical perspective suggests very much the latter. On this and other issues relating to health inequalities, the paper by Arcaya, Arcaya and Subramanian (2015) is valuable.

There is a vast literature on health inequalities in countries of the Global North, to which geographers have made key contributions, such as in the United States (McLafferty, Wang & Butler 2011), Canada (Shantz & Elliott 2021), the United Kingdom (Bambra 2016; Dorling 2013), New Zealand (Pearce, Tisch & Barnett 2008) and elsewhere. I do not intend to review these contributions, preferring to focus attention on some countries of the Global South. However, within some countries of the Global North there are marginalized groups to which I do want to give attention; in particular, as we see later, there are Indigenous populations occupying places that are as neglected as those who inhabit them.

The Global North literature on health inequalities tends to focus both on regional inequalities and, frequently, on more local – neighbourhood – variations. A classic distinction in this literature is between compositional and contextual effects (Gatrell & Elliott 2015: 82; Brown *et al.* 2018: ch. 8). In other words, is poor health in specific settings determined by the fact that poor people, with health-damaging behaviours, live there (the "composition" of a place) or by the fact that people live in places ("contexts") that are environmentally or socio-economically deprived? Writers such as Bambra (2016) offer a more relational view, suggesting that the behaviours of individuals are shaped by the features of the areas in which they live; context and composition are not mutually exclusive.

Bambra, Smith and Pearce (2019) have called for a political economy approach to the geography of health inequalities, arguing that the context–composition debate has deflected attention away from the influences of macro-political and economic determinants of health. National government policy shapes or constrains what can be done regionally and locally; thus, for example, the austerity measures introduced in the United Kingdom and Europe in recent years have contributed to poor health and health inequalities (Stuckler *et al.* 2017; Box 2.1).

The literature on health inequalities owes much to classic work by Göran Dahlgren and Margaret Whitehead (who have recently reviewed their contribution: see Dahlgren & Whitehead 2021), although their original work was on the determinants of health, not the determinants of health inequalities. For example, although having clean water and sanitation are determinants of health (water, sanitation and hygiene are often referred to as "WaSH"), it is differential exposure or access to these that leads to inequalities. The same is true of work: having a job may be good for health, but different jobs produce different exposures to things (pollution, hazards, and so on) that cause gradients of inequality. Who gets clean water, good jobs and better healthcare is socially determined, and the "social determinants of health" agenda has spawned a huge literature. But, alongside accepting social determinants, I would add that the question as to who gets what *and where* is both socially *and* politically determined. Determinants, including where services are provided, and to whom, are geographical as well as social. It is therefore important to acknowledge what a geographical perspective on health inequalities offers – which means looking at where the goods and "bads" are distributed, and their health consequences. In exploring social determinants, we need to know where the poor are living, where incomes are low, where water and air quality are poor, where food is inadequate, and so on. Of course, the explanations may well lie in politics and power relationships; but these play out differently in different places.

When speaking of "determinants", we need to acknowledge that these are political and commercial, not solely social (Box 2.1).

A counter to the literature on determinants is provided by Duff (2018: 140), who has questioned "the whole notion of social determinants as being objective, stable, discrete realities". Andrews (2021) concurs: "[A]t issue here is the reliance on a creeping agenda which increasingly loses touch with life's vitality as ever more social structures, contexts and behaviours are incorporated as determinants, each reduced and abstracted further

BOX 2.1 POLITICAL AND COMMERCIAL DETERMINANTS OF HEALTH

Ilona Kickbusch (2005) introduced the notion of political determinants of health, a theme she revisited ten years later (Kickbusch 2015). By "political determinants" she means considering how power, whether vested internationally, nationally, regionally or locally, affects health outcomes. Decisions that are taken, for example, about regional investment can potentially boost local economies, providing employment, which is clearly beneficial for health. Following the international recession in 2007, the European Commission, along with the European Central Bank and International Monetary Fund (IMF), prioritized reductions in public expenditure in order to reduce budget deficits. Such austerity measures have been shown to be regressive (Stuckler et al. 2017), with measurable impacts on mental health and suicide. Austerity measures are a political choice.

Decisions that governments take about international aid affect the health of those living in the Global South, although such decisions need critical scrutiny. Issues of governance are considered more fully in Chapter 4, while many of the inequities considered in this chapter result from political decisions taken at different levels.

Although the literature on political determinants of health pays considerable attention to the economic and structural power exercised by governments and international organizations, it seems to have less to say about the physical power exercised by one group on another. For example, one country can control access to water resources needed by its neighbour. Kurdish people living in northeast Syria complain that the Turkish authorities are denying them access to water from the river Euphrates that they use to irrigate fields – leading to crop failure and malnutrition – although climate change (see Chapter 9) may also play a part in these water shortages (Laffert & Sala 2021).

"Commercial determinants" of health relate to the power exercised by large corporations: the strategies they use to promote the consumption of products (such as tobacco or unhealthy foods) that harm health. These corporations may seek to negate or overturn policies adopted by governments to protect public health. The interests of energy companies – gas and oil – and the countries that rely on them are threatened by policies to reduce global warming. For example, in the run-up to COP26 (the 26th United Nations (UN) Climate Change Conference, held in Glasgow in 2021) it was revealed in leaked documents that Saudi Arabia's oil ministry had demanded that phrases such as "the need for urgent and accelerated mitigation actions at all scales ... " should be eliminated from a UN report. In addition, a senior Australian government official reportedly rejected the conclusion that it was necessary to close down coal-fired power plants (BBC News 2021c).

Soft power can be exercised to shape preferences; for example, tobacco companies can encourage smokers not to quit but, rather, to purchase "vape" products instead.

Clearly, commercial determinants of health play a major role in the advance of non-communicable diseases. In addition, the Canadian Public Health Association (CPHA) also refers to "ecological determinants" of health: the natural environmental goods (such as water, air and soil) that affect human health (CPHA 2015). These are considered in more detail in Chapter 7.

For further reading on commercial determinants, see Kickbusch, Allen and Franz (2016) and papers referenced in de Lacy-Vawdon and Livingstone (2020), while Crocetti et al. (2022) review these in connection with Indigenous health. On political determinants, the paper by Moon (2019) is an important source, as is King (2017).

---

and further to statistical levels." As we saw earlier, such poststructuralist perspectives are gaining momentum within health geography. Nonetheless, measuring, mapping and analysing such determinants and the resulting health inequities remain of considerable importance, certainly in the Global South.

Before considering in more detail such inequities and their determinants in the Global South I want to consider the Sustainable Development Goals, the successor to the Millennium Development Goals. I then go on to look at some regional "canvases" – inequalities as manifested in selected countries.

Sustainable Development Goals

Although this is not a book about sustainable development, that agenda provides a suitable way in to the topic of inequality within global health geography.

In 2015 the United Nations General Assembly agreed to establish Sustainable Development Goals (SDGs) – 17 in total (Table 2.1) – the set of which forms a "blueprint to achieve a better and more sustainable future for all" (UN n.d.a), with a target date for achievement of 2030. Within the set of 17 goals are 169 targets and 231 indicators. Clearly, the goals are interrelated, and each one – and many targets therein – has direct or indirect potential impacts on health. For example, under Goal 3 ("Ensure healthy lives … "), one target is to "end the epidemics of AIDS, tuberculosis, malaria and neglected tropical diseases and combat hepatitis, waterborne diseases and other communicable diseases", and specific measurable indicators are incidence rates of these diseases. Further, since global health demands an engagement with inequality, it is evident that several goals (for example, 4, 5 and particularly 10) are quite specific on this.

Laudable though the aims of the SDGs are, many are so obviously unachievable during a 15-year lifetime (for example: "End poverty in all its forms everywhere"; "Promote

Table 2.1 Sustainable Development Goals

1	End poverty in all its forms everywhere.
2	End hunger, achieve food security and improved nutrition and promote sustainable agriculture.
3	Ensure healthy lives and promote well-being for all at all ages.
4	Ensure inclusive and equitable quality education and promote lifelong learning opportunities for all.
5	Achieve gender equality and empower all women and girls.
6	Ensure availability and sustainable management of water and sanitation for all.
7	Ensure access to affordable, reliable, sustainable and modern energy for all.
8	Promote sustained, inclusive and sustainable economic growth, full and productive employment and decent work for all.
9	Build resilient infrastructure, promote inclusive and sustainable industrialization and foster innovation.
10	Reduce inequality within and among countries.
11	Make cities and human settlements inclusive, safe, resilient and sustainable.
12	Ensure sustainable consumption and production patterns.
13	Take urgent action to combat climate change and its impact.
14	Conserve and sustainably use the oceans, seas and marine resources for sustainable development.
15	Protect, restore and promote sustainable use of terrestrial ecosystems, sustainably manage forests, combat desertification, and halt and reverse land degradation and halt biodiversity loss.
16	Promote peaceful and inclusive societies for sustainable development, provide access to justice for all and build effective, accountable and inclusive institutions at all levels.
17	Strengthen the means of implementation and revitalize the Global Partnership for Sustainable Development.

Source: UN (n.d.a).

peaceful and inclusive societies"), not to mention a far more extended period of time, that hard questions need to be asked about their purpose and intent. Further, many targets and indicators are hard to measure and then monitor, even assuming data are available from the poorest countries. Consequently, it is unsurprising that the UN's SDGs have come in for serious criticism. For example, in a brief polemic, Easterly (2015: 322) suggests that "the SDGs are about as likely to result in progress as beauty pageant contestants' calls for world peace".

Moreover, one can argue that the SDGs derive from a neoliberal perspective that ignores entrenched power relations; the root causes of poverty – itself the major cause of ill health and mortality – go unexamined. As one feminist critic puts it:

> [The SDG agenda] does not attempt to transform power relations between the North and the South, between the rich and the poor, and between men and women. Agenda 2030 [which established the SDGs] aims at 'transforming our world', but intends to get there without substantially opposing the powers that be. (Esquivel 2016: 11)

Agenda 2030 made no reference to pandemics. In 2015 there was optimism about the prospects for international cooperation and economic growth, even if the benefits of such growth were unlikely to be shared out. As we shall see in Chapter 8, Covid-19 has completely altered the narrative, driving very large nails into what is likely to be the SDG coffin. Intergovernmental cooperation (for example, over vaccine supply) has been found wanting and aid for the poorest countries is set to fall. The British government, for example, has reneged on its manifesto commitment of 0.7 per cent of national income, reducing this to 0.5 per cent (a cut of £4 billion a year: BBC News 2021e). A recent editorial in the journal *Nature* (2020) makes the point very starkly, suggesting that "the coronavirus pandemic has put the Sustainable Development Goals (SDGs) out of reach. Most of the goals to end poverty, protect the environment and support well-being by 2030 were already off course. Now, what little progress had been made has been stopped in its tracks."

As we see from Table 2.1, SDG 1 aims to "[e]nd poverty in all its forms everywhere". The World Population Review says that, in 2021,

> While poverty rates have made significant improvements over the past few decades, 736 million people live in extreme poverty, surviving on less than $1.90 per day [the World Bank's international poverty line]. 413 million of these people live in sub-Saharan Africa. Nineteen countries worldwide have poverty rates over 50 per cent. (World Population Review n.d.)

Inevitably, Covid-19 has drastically altered these estimates: "About 120 million *additional* people are living in poverty as a result of the pandemic, with the total expected to rise to about 150 million by the end of 2021" (World Bank 2022, emphasis added). The Bill & Melinda Gates Foundation reports annually on progress in meeting SDG targets (see, for example, Gates Foundation 2022).

The World Inequalities Database (WID), led by French economist Thomas Piketty, has produced data on inequality for 173 countries. The website reports that

> Latin America and the Middle East stand as the world's most unequal regions, with the top 10 per cent of the income distribution capturing respectively 54 per cent and 56 per cent of the average national income. In Latin America, amidst a decline of inequality levels in a handful of countries, inequality persisted, and even increased, in some others. (WID 2020)

In India, the top 10 per cent income share grew from 30 per cent in the 1980s to over 56 per cent in 2020, following deregulation and liberalization reforms. The WID also notes that "extreme inequality levels can be found among nations which historically experienced white settlers' colonization and extreme forms of racial injustices (e.g. South Africa)" (WID 2020). The relation between income inequality and health is explored in the next section.

Income inequality and health

Kawachi and Wamala (2006b: 125) assert that, "[o]ther things being equal, a country or region with greater inequality in the distribution of incomes will have worse average health status than a country or region with a more egalitarian distribution of incomes". The relationship between income inequality and health has been popularized by Wilkinson and Pickett (2010), who produce a wealth of evidence in favour of the association.

Dorling, Mitchell and Pearce (2007) have examined that association using data for 126 countries but, crucially, looking at the association for different age groups. They find a clear and statistically significant relationship between all-cause mortality and income inequality (measured by the Gini index), a relationship that is strongest at ages of between 25 and 39 years. The relation is strong for both sexes but slightly more so for males. Ward and Viner (2017) build on this work by looking at the association between income inequality and child and adolescent mortality among a set of 103 low- and middle-income countries. After standardizing for national wealth, they find that inequality (also measured using the Gini index) is positively associated with all-cause and communicable disease mortality, among both boys and girls, and across all age groups. The association between income inequality and mortality strengthens as age increases. They conclude that there needs to be a focus on "upstream" factors – policies to reduce income inequality – if there are to be significant reductions in such mortality.

Detailed research on the association between income inequality and health outcomes has been undertaken by several scholars working in Brazil. Vincens, Emmelin and Stafström (2018) use individual and neighbourhood data to show that income equality and neighbourhood infrastructure work in tandem to affect self-rated health. Individuals living in more egalitarian states appear to have even better health if they also live in neighbourhoods with better-quality infrastructure (basic services such as water, sewage, electricity and waste collection). Income inequality is highest in Amazonas, Rio de Janeiro and São Paulo states (see Figure 2.2).

Vincens and Stafström (2015) explore the relation between income inequality and deaths from stroke in Brazil. Their results show that income inequality is associated with stroke mortality; a reduction of ten points in the Gini index (in other words, improved levels of income equality) is associated with an 18 per cent decline in stroke mortality. Gaspar *et al.* (2021) have extended this research in Brazil to look at associations with mortality and morbidity from many non-communicable diseases (NCDs). Although they find no association between income inequality and the overall burden of all NCDs, they do find that income inequality is linked to higher burdens from diabetes and from alcohol-related diseases, particularly for men. Correa Massa and Chiavegatto Filho (2021) report on the association between income inequality and self-reported health in older people living in the 27 state capitals of Brazil. They find evidence of poorer self-reported health among older adults living in cities of above-average income inequality, suggesting that living in unequal areas can have a detrimental impact on the health of older adults.

This association is independent of individual factors such as education attainment and reported chronic illness, supporting (as have other studies cited here) the Wilkinson–Pickett arguments that income inequality can directly affect individual health. I consider Brazil in more detail below.

Adjaye-Gbewonyo *et al.* (2016) sound a note of caution in this kind of work. They suggest that, in countries where many are living at or below the poverty line, health is determined more by absolute income than by relative income levels; income inequality has more of an impact in high-income countries, where basic needs are, in general, already catered for.

Health inequalities in the Global South: the big picture

An important source of information on global health is provided by the Global Burden of Disease (GBD) programme, a major international collaboration based initially at Harvard University in the 1990s, now at the University of Washington (Seattle). Funded by the Bill & Melinda Gates Foundation, it has a close association with the World Health Organization (WHO) and the World Bank. Its overarching aim is to develop, and provide, reliable epidemiological information on a wide range of diseases, injuries and risk factors in order to provide international comparisons and policy advice. Its body of evidence is considerable (see Institute for Health Metrics and Evaluation [IHME] 2020).

The GBD 2019 Demographics Collaborators (2020) provide data on under-five mortality, life expectancy (male and female) and healthy life expectancy for 204 countries. The research reveals that global life expectancy at birth has increased from 51.1 years in 1950 to 67.2 years in 2000 and 73.5 years in 2019. But this of course masks considerable variation. Their paper can be consulted for data on life expectancy for specific countries; data for countries in sub-Saharan Africa, where it was lowest in 2019, are given in Table 2.2.

Table 2.2 Life expectancy in selected sub-Saharan countries, 2019

	Female life expectancy at birth	*Male life expectancy at birth*
Lesotho	55.4	48.6
Central African Republic	55.8	49.3
Somalia	61.3	55.8
Chad	62.2	58.9
Sierra Leone	63.1	60.6
Guinea-Bissau	64.0	58.1
Nigeria	65.9	62.8
South Sudan	66.2	61.7
Côte d'Ivoire	67.0	62.0
Equatorial Guinea	67.1	64.5

Source: GBD 2019 Demographics Collaborators (2020).

Other GBD research looks at specific population groups. For example, the GBD 2019 Adolescent Mortality Collaborators (2021) present findings on mortality among ten- to 24-year-olds. About one-third of deaths among this group are found to result from transport injuries, other injuries or interpersonal violence or conflict; a further third are attributable to communicable disease or poor nutrition; 27 per cent are the result of non-communicable diseases; and about 8 per cent are caused by self-harm. But these proportions vary among major world regions; for example, in Latin America and the Caribbean the leading cause of death is interpersonal violence and conflict.

The GBD provides a valuable source of descriptive data that permits international comparisons – a point acknowledged even by its critics, such as Jeremy Shiffman and Yusra Shawar (2020). These writers have been particularly vocal about what they term the global health metrics "enterprise", demanding that its political voice and power (concentrated in the Global North and among elite interest groups), as well as its key role in global health governance, be acknowledged and scrutinized. They argue that complex health issues are reduced to quantifiable outputs; for example, breastfeeding rates are easier to measure than a child's right to play. But their main concern is that data and power are transferred from low-income countries to institutions in high-income countries, ensuring that objectives set by the latter, and global organizations, are met. In a telling statement, of considerable relevance to geographers, they write:

> The reason the GBD map – or any other map that global health metrics producers might offer of the world – receives greater and greater acceptance has not so much to do with its scientific superiority (however that term is understood) but with the resources, social connections, and power of its primary backers.
> (Shiffman & Shawar 2020: 1454)

This of course links back to the issue of decolonization, while Paul Farmer has observed that "the experience of suffering … is not effectively conveyed by statistics or graphs" (Farmer 2005: 31). Yet, despite these strictures, some geographical research on variations in health indicators proves useful in mapping where priority needs to be given to addressing the worst inequity.

There are many examples of health inequalities, only some of which are considered here. The main focus is on reproductive, maternal, newborn and child health (abbreviated to RMNCH). A useful resource is the WHO's Health Inequality Monitor (see www.who. int/data/inequality-monitor), which provides up-to-date data (including some valuable interactive visualizations) for a variety of RMNCH indicators, based on a large number of household surveys undertaken in over 100 countries between 1991 and 2019.

Detailed examples are provided by research in sub-Saharan Africa. Yourkavitch *et al.* (2018) map several child health indicators using demographic and health survey data for 27 countries obtained between 2010 and 2014. All show clear within-country variability, such as in under-five mortality. Between countries, the rate ranges from 52 per 1,000 live births in Kenya to 156 per 1,000 in Sierra Leone but there is clear subnational variation, with Burkina Faso having the greatest range (75 to 207 per 1,000). The rates are no respecter of boundaries; rates are high across the Sierra Leone/Guinea/Côte d'Ivoire borders, and in northern Nigeria, Cameroon and Burkina Faso. Many ethnic groups live in areas that cut across the boundaries drawn under colonization; for example, the nineteenth-century (Sunni Muslim) Sokoto Caliphate encompassed Nigeria, Niger, Cameroon and Burkina Faso before it was abolished during British (and, in the case of Cameroon, German) control in 1903.

Faye *et al.* (2020) pursue this further, confirming the extent of geographical disparities in RMNCH indicators within countries of sub-Saharan Africa. Most countries with high inequality are found in west and central Africa, but with considerable variation between countries within the subregion. Cameroon, Nigeria and Mauritania have the most unequal subnational coverage, followed by Angola, the Central African Republic (CAR) and Ethiopia. Many of these have seen recent or ongoing conflict and violence. In contrast, countries such as Rwanda and Malawi had better RMNCH indicators, as well as lower levels of within-country inequality.

Urban–rural divides matter too. As revealed by Victora *et al.* (2019) in their study of LMICs, inequalities in the proportion of births taking place in health facilities ("institutional delivery") varies from country to country but also by whether the birth is in an urban or rural area (Figure 2.1) and according to wealth (family assets). Unsurprisingly,

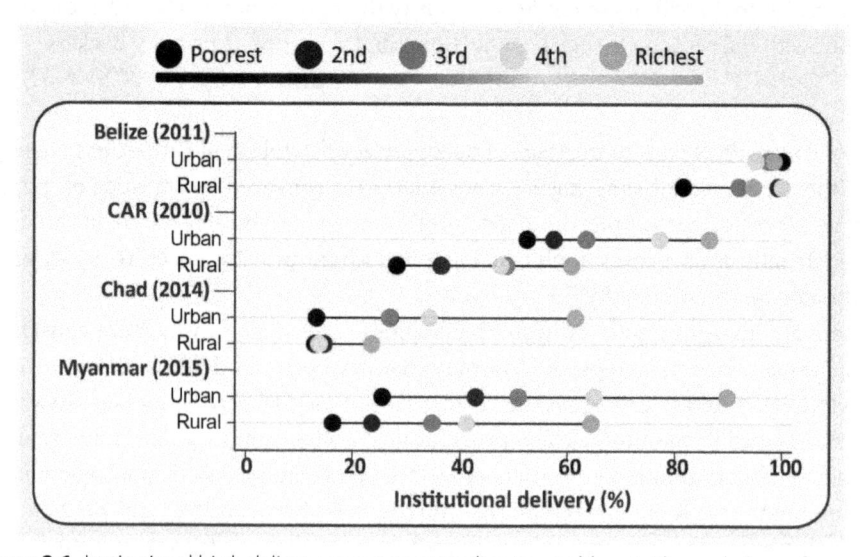

Figure 2.1 Institutional birth delivery coverage according to wealth quintiles and place of residence in selected countries
Source: Victora *et al.* (2019).

the proportions of health centre births are higher in urban than rural areas, but there are varying inequalities within both urban and rural areas according to wealth. For example, in Chad there is clear inequality between urban areas, but in rural areas, although there is low overall coverage, there are few wealth-based inequalities. In Myanmar, both urban and rural areas reveal large inequalities, with hospital births to women in rural areas lagging well behind urban women, regardless of wealth status.

One of the targets within the all-encompassing SDG 3 is to improve levels of childhood immunization, including that against measles. Although measles cases declined globally between 2010 and 2016, they have increased again since then. Vaccination programmes have been disrupted by Covid-19, and this will inevitably lead to further morbidity and mortality. There is a clear geography to immunization coverage, as revealed by a group reporting in *Nature* (Local Burden of Disease Vaccine Coverage Collaborators 2021). The article demonstrates significant between-country and within-country estimates of vaccine coverage. Despite improvements in vaccine coverage between 2000 and 2019, progress has stalled in the last ten years. Data for 2020 indicate that the percentage of one-year-old children who had received their first measles vaccination stood at 44 per cent in Angola, 54 per cent in Nigeria, 59 per cent in Madagascar and 60 per cent in Ethiopia, but about 80 per cent in Namibia and Ghana (WHO n.d.a). However, there are clear regional differences in these five countries.

In general, coverage in 2020 was lower in remote rural areas, where an estimated 33 per cent of children were unvaccinated, compared with 15 per cent of children living in urban areas. Economic instability and political conflict in some countries work against vaccination coverage. For example, in Angola, such instability led to a 28 per cent reduction in government health spending per capita between 2010 and 2018, and this may have contributed to declines in vaccination coverage. Lower rates in northern Nigeria than in southern states are also likely to have resulted from political conflict in border regions.

Health inequalities: regional canvases

Having painted quite a broad-brush picture using some RMNCH health indicators, detailed attention is given to five countries, each drawn from a category of the World Bank's 2021 classification by income. These are Brazil and China (upper middle income), India (lower middle income) and Mozambique and the Democratic Republic of the Congo (DRC: low income). Of these, two have a history that reflects colonization by Portugal, and one (the DRC) by Belgium, while India was under direct British rule from 1858 until independence in 1947. These country-specific portraits – inevitably brief, of course – help make the point that, just as there are regional health divides in countries of the Global North, so too it is absurd to portray large countries such as India, China and Brazil as homogeneous spaces.

Brazil

Brazil is a middle-income country (population 213 million in 2020), divided politically into five distinct regions (North, Northeast, Southeast, South and Central-West; see Figure 2.2). These five regions are subdivided into 27 states. The Southeast comprises about 40 per cent of the population, including the most populous states of São Paulo and Rio de Janeiro. Like the South (which comprises 14 per cent of the total population), it is one of the two wealthiest regions.

Brazil's public healthcare system, introduced in 1988, is known as SUS (Sistema Único de Saúde: Unified Health System). Part of this involved creating a Family Health Strategy, comprising primary care workers who are responsible for small neighbourhoods. Evidence suggests that, between 1990 and 2014, there were significant improvements in overall population health, and a reduction in health inequalities. Data on mortality rates for children aged under five years reveal that there was a fivefold gap between states with the highest and the lowest levels in 1990 but this had reduced to a 2.5-fold difference by 2014 (Machado & Silva 2019; Machado *et al.* 2020).

Between 2016 and 2022 Brazil was led by conservative governments (President Temer 2016–18, President Bolsonaro 2019–22) that introduced austerity measures

Figure 2.2 Brazil: regions and states
Source: Galvão de Araújo *et al.* (2012).

and regressive social reforms. "In only a short time, it has been possible to observe worsening social indicators, such as rates of poverty and extreme poverty, along with stagnation in the reduction of social inequalities that had occurred between 1990 and 2014" (Machado & Silva 2019). Castro *et al.* (2019) warn that fiscal policies implemented in 2016 by Michel Temer brought austerity measures that threatened the prospect of the SUS in providing universal healthcare. Daly (2019) asserts that the election of Jair Bolsonaro marked "not the beginning of a democratic crisis for Brazil, but the punctuation and intensification of a process of decay that has affected the country's democratic system for some time". This process included the appointment of military figures to key government positions.

Szwarcwald *et al.* (2016) have examined broad regional variations in life expectancy in Brazil, using data from the national health survey conducted in 2013. They report clear inequalities in health status and some of the determinants thereof (such as education, socio-economic class and access to healthcare; see Table 2.3). The worst health, and the disadvantages that contribute to this, are in the North and Northeast regions of the country. The proportion of poor self-rated health in Brazil as a whole was 5.8 per cent but this varied from 4.4 per cent in the Southeast to 8.3 per cent in the Northeast. Those living in the North and Northeast have, on average, six years' lower life expectancy at age 20 than those in the South and Southeast. Rasella *et al.* (2016) reveal considerable geographical detail on poverty, illiteracy and life expectancy; see Figure 2.3.

It is important to acknowledge, here and elsewhere in this book, that pictures of health and its determinants can change dramatically within a few years. For example, although health inequalities in Brazil improved in the first decade of the twenty-first century, they have worsened subsequently. Moreover, Bolsonaro's government set the country back in terms of human rights. The Brazil pages of the 2021 *World Report* by Human Rights Watch (HRW 2021) report that environmental law was weakened, enabling criminal networks that undertake illegal deforestation in the Amazon and threaten those protecting the rainforest. Bolsonaro referred to NGOs working in the Amazon as

Table 2.3 Health and social indicators in regions in Brazil, 2013

	North	Northeast	Southeast	South	Central-West
Healthy life expectancy at age 20 (years)	48.5	48.7	54.9	54.1	52.1
Completing secondary school education (per cent)	41.7	37.7	50.9	45.5	47.4
Lowest socio-economic class (per cent)	40.9	42.5	15.1	13.6	21.5
Consulting a doctor within last 12 months (per cent)	66.4	69.7	79.5	78.6	73.8
Private health insurance (per cent)	12.6	14.7	34.5	31.4	28.0

Source: Szwarcwald *et al.* (2016).

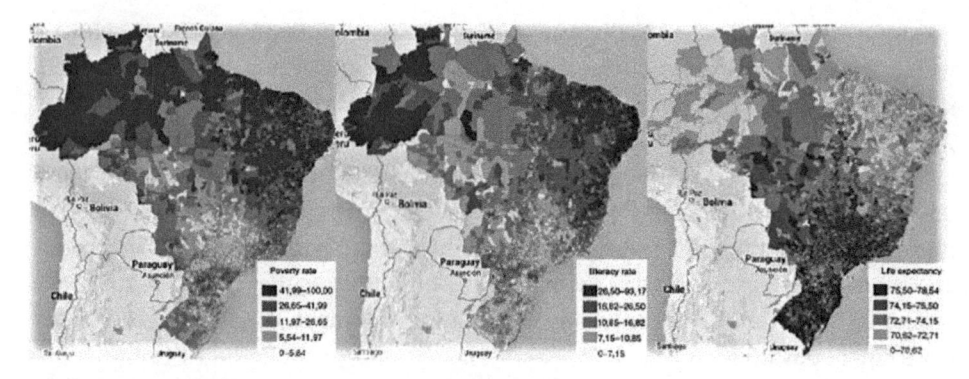

Figure 2.3 Poverty, illiteracy and life expectancy in Brazil
Source: Rasella *et al.* (2016).

a "cancer" that he "can't kill", in addition to blaming Indigenous communities for fires in the Amazon rainforest that are causing air pollution. Further, extra-judicial killings increased: in Rio de Janeiro, between January and May 2020 police killed 744 people, the highest number since 2003, despite overall crime levels having reduced.

In a devastating conclusion to their study, Reichenheim *et al.* (2011: 1971) state: "Brazil has always been a violent country: national development began with the enslavement of Indians and Black Africans, and the scars of the country's colonial past persist to this day." They assert that corruption, poverty and inequality persist, with the government failing to protect human rights and provide basic services.

India

India's population was estimated in 2021 as 1.34 billion – comprising about a sixth of the global population – making it the second most populous country in the world. Part of the British Empire from 1858 (the so-called British Raj), it gained independence in 1947. The country is divided into 28 states and eight union territories (Figure 2.4). There are considerable regional differences in terms of economy and society. Although particular attention is given here to health outcomes, it is evident that the service sector has driven economic growth more than manufacturing, and that highly qualified Indians gravitate to cities (such as Bangalore, Hyderabad and Mumbai) where the high-tech firms are located.

The impact of British rule in India is a contested subject. The country was a source of raw materials (e.g. cotton and jute) in the nineteenth century that helped generate Britain's Industrial Revolution, and in return India became a market for the goods that were manufactured in Britain. The infrastructure (notably, the rail network) that Britain oversaw enabled the transport of raw materials and finished products, but it can be seen as part of an extractive economy. Britain's intervention in Indian agriculture led to a succession of famines, killing perhaps 8 million during the 1876–88 drought; nonetheless, despite crop failures exports of grain to Britain continued.

Amrith (2009) has pointed out that a further legacy of British rule was one of underinvestment in health and that public health was given relatively little attention after independence. Attention was given more to "vertical" programmes, such as malaria control, at the expense of more "horizontal" programmes of improving sanitation. Some states, such as Kerala and Tamil Nadu, gave broader public health measures, including childhood immunization and improving malnutrition, more consideration.

In 2015 there were just over 25 million live births but 1.2 million under-five deaths, a mortality rate of 48 per 1,000 live births, which marked a decline from 119 per 1,000 live births in 1992. But that national rate masks considerable variation by caste and socio-economic status – as well as by region and state. Half the under-five deaths in 2015 were in just three states in northern India: Bihar, Madhya Pradesh and Uttar Pradesh. In these states the leading causes of death (pneumonia, diarrhoea) were related to infectious disease spread, while in states with low mortality (Table 2.4) the leading causes were all non-communicable (such as pre-term birth complications and birth defects: Liu *et al.* 2019). Of course, these states cover vast areas, and it is important to recognize that there will be significant intra-state variations. Puri *et al.* (2020) have therefore undertaken a more detailed spatial analysis of child health, using 2015–16 National Family Health Survey data disaggregated to 640 districts. Their research shows both regional and local variation in rates of immunization, anaemia and stunting. Among the factors explaining these inequalities, they suggest that access to, and use of, health facilities is hindered by malpractice and corruption, but poverty and female literacy (Table 2.4) are associated with poor child health (although it is worth noting that it is the male heads of household who are the main decision-makers). In addition, children of Muslim parents, and those in the Scheduled Tribe or Scheduled Caste groups, have worse health outcomes.

Table 2.4 States in India with health indicators

State	Under-five mortality rate (per 1,000 live births), 2015	Percentage of persons below poverty line (2011–12)	Life expectancy at birth (2020)	Female literacy rate (2011)	Estimated risk of death between 30 and 70 years per 1,000 (2017)	
					Men	Women
Bihar	52.42	33.74	68.1	53.33	419	330
Madhya Pradesh	67.07	31.65	64.2	60.02	448	306
Uttar Pradesh	61.15	29.43	64.1	59.26	471	373
Tamil Nadu	21.71	11.28	70.6	73.86	442	287
Kerala	12.50	7.05	74.9	91.98	351	199
Maharashtra	24.07	17.35	71.6	75.48	354	279

Sources: Liu *et al.* (2019), Jose (2019) and Rao *et al.* (2020).

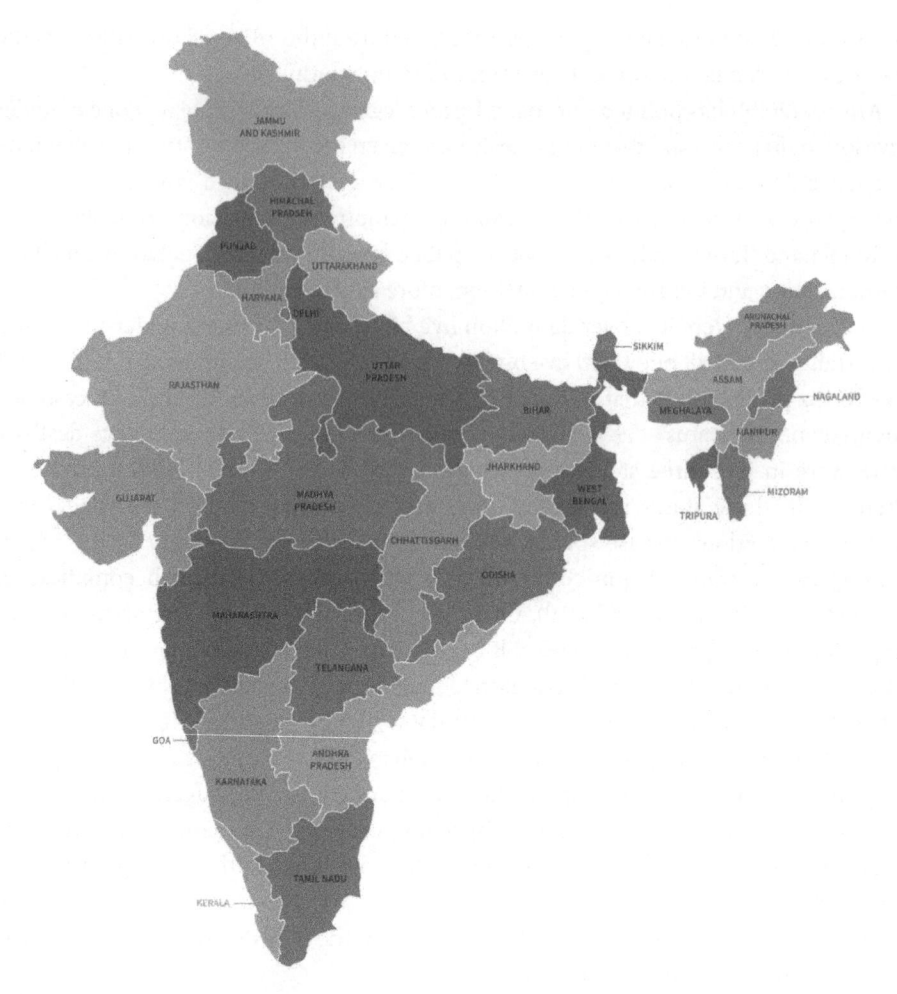

Figure 2.4 India: states and territories
Source: iStock.

The percentages of poor households, illiteracy and home births are higher in the centre and east of the country, compared to the south (Figure 2.4; Table 2.4). All these figures have consequences for achieving the SDG child survival targets. Liu *et al.* (2019) suggest that, to achieve this by 2030, ten of the 25 states will need to increase their annual rate of reduction. As Table 2.4 shows, such regional inequalities are also apparent among adults, the premature death rates being a function of high rates of non-communicable diseases as well as tuberculosis (TB) (Rao *et al.* 2020).

Crabtree (2018) reports that, since Narendra Modi came to power as prime minister with his centre-right Bharatiya Janata Party (BJP), "the spoils of India's increasing integration with the global economy have been enjoyed disproportionately by the already-prosperous". The richest tenth of Indians hold nearly three-quarters of its wealth. Although some economists applaud higher overall standards of living, investment in infrastructure and export-focused manufacturing, others (such as Nobel laureate Amartya Sen) say that growth has come at the cost of social progress among the poor. Sen is critical of India's neoliberalism and would prefer to see more resources directed at education and health, which he considers have been neglected by Modi's government. Crabtree (2018) suggests that more tax revenue needs to be collected from wealthier citizens, a call that many make in some countries in the Global North.

China

During a period of rapid economic growth in China since 1990 various health indicators have shown significant improvement, albeit accompanied by rising income inequality. Wang *et al.* (2016) have provided estimates of under-five mortality for the 31 provinces (Figure 2.5) and 2,851 counties in mainland China. Although the overall mortality rate

Figure 2.5 China: provinces
Source: iStock.

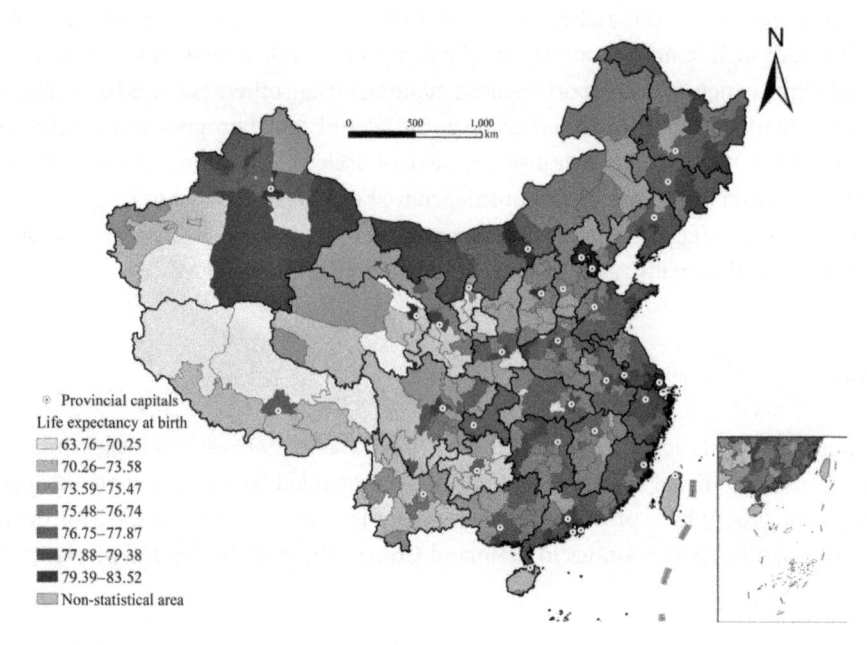

Figure 2.6 Life expectancy at birth in China, 2010
Source: Tao *et al.* (2021) (courtesy of Dr Huang Daquan, Beijing Normal University).

declined from 46 per 1,000 live births in 1990 to 13.7 in 2008, the rate varied from 3.3 per 1,000 livebirths in one county (Huangpu), in Shanghai, to 104.4 per 1,000 live births in the county of Zamtang in Sichuan province (see Wang *et al.* 2016 for a detailed map). As the authors note, such variation is "astounding", and they suggest that the highest rate is no different from Burkina Faso in sub-Saharan Africa.

Globalization has clearly benefited coastal regions of China, with inland regions lagging behind in per capita gross domestic product (GDP). Such regional divisions are mirrored in many health outcomes. For example, life expectancy is much greater in the east, lower in central China, and worst in the west and northwest (Figure 2.6). Luo and Xie (2020) link variations in life expectancy to income inequality, arguing that China's rapid economic growth in recent years has been paralleled by rising income inequality; between 1978 and 2012 the Gini index increased from 0.28 to 0.56.

In a recent overview of progress towards meeting SDG targets, Chen *et al.* (2019) point to the major challenge of reducing regional health inequalities. They too confirm the overall pattern shown in Figure 2.6: in general, provinces in the east show much better health outcomes and access to care than central and western provinces. They suggest that two-thirds of China's provinces will not achieve half the 28 health-related SDG indicators by 2030.

Health inequalities are, of course, not only regional; they are ethnic too, as suggested in Box 2.2.

BOX 2.2 THE UIGHUR POPULATION IN XINJIANG PROVINCE

In recent years the treatment of a minority Muslim population, the Uighurs (sometimes spelt "Uyghurs") in northwest China, has come to wider attention. The Uighurs number about 12 million in the Xinjiang autonomous region. Some estimates suggest that up to 1 million Uighurs are detained in so-called "re-education" camps, designed to ensure that those incarcerated learn to adhere to Communist Party ideology. Officially, the Chinese government sees these as a means of combating terrorism and Islamist extremism (BBC News 2021b). Human rights groups have accused the government of forced labour (especially in the region's important cotton fields) and the forced sterilization of Uighur women, leading to accusations of genocide (defined by the UN as "intent to destroy, in whole or in part, a national, ethnical, racial or religious group": UN n.d.b).

There are clear differences in the health outcomes of the Uighurs and the majority Han Chinese. Using official Chinese data for 2000–04, Schuster (2009) cites life expectancy as 63 years for the Uighurs and 73 years for the Han population. She quotes an infant mortality rate for Han in Xinjiang of 13.1 per 1,000 births, but 101.7 per 1,000 for Uighurs. For the Uighurs there is a clear gradient in infant mortality from town to the most remote rural areas, but little evidence of such for the Han. Her analysis suggests that infant mortality rates are dependent on household income and parental employment and education, but her overall conclusion is that poor health outcomes among the Uighurs are associated very clearly with their ethnicity.

As a major world power, China is seeking to extend its sphere of influence well away from east Asia, as instanced by its Belt and Road Initiative (BRI), which builds on historic trade routes (the former "Silk Road") across the rest of the continent. This is seen by some commentators, such as Hillary Clinton, as a "neo-colonialist" project (Krause-Jackson 2011), not least given China's ambitions away from Asia, in Africa. Other writers, such as Shah (2020), point to the continuous entanglement of major Western powers in African affairs and suggest we "avoid falling for the widespread myth that there exists a binary divide between altruistic western intervention in Africa and exploitative Chinese interference". Within the BRI is "a substantial health component" (Zhou *et al.* 2019: 1146), with the WHO referring to a "health Silk Road", and the Chinese president, Xi Jinping, linked this to the production and distribution of Covid-19 vaccines. Before the pandemic arrived, a Forum on China–Africa Cooperation had been created, involving the provision of medical supplies, programmes for training doctors, and the construction of new hospitals. Nonetheless, other priorities lie more in massive infrastructure projects, including port development and rail expansion in east Africa.

Mozambique

Mozambique is divided administratively into ten provinces (and, further, into 129 districts); see Figure 2.7. Its population has risen sharply since 2010, from 23.5 million to 31.3 million in 2020, a growth rate of 33 per cent. It gained independence from Portugal in 1975, but conflict between the two leading independence groups (FRELIMO and RENAMO) led, until 1992, to over 1 million deaths and massive population displacement, either to neighbouring countries or to major cities such as Maputo and Beira.

The country's northern borders with Tanzania are porous. Recognizing the incursions from jihadist group Ansar al-Sunnah, Will Marshall (2021), from Global Risk Insights, writes: "[T]he key takeaway for policymakers seeking to understand contemporary conflict dynamics in northern Mozambique should be a contextual appreciation of the complex political economy of [the northeastern province of] Cabo Delgado which underpins localised grievances against both the state and foreign investors."

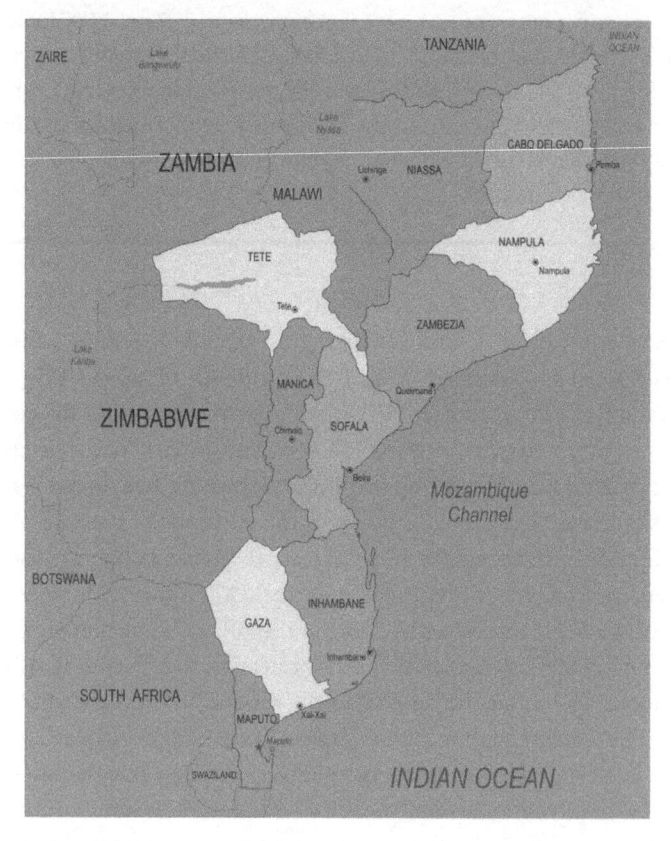

Figure 2.7 Provinces of Mozambique
Source: iStock.

Silva (2008) notes that the country took on an IMF structural adjustment programme in 1987, which led to the closure of state-run factories, such privatization benefiting mainly foreign-owned multinational corporations. Mugabe *et al.* (2021) note that in the late 1990s Mozambique experienced a high rate of economic growth, accompanied by increased average household income and improved infrastructure (water, sanitation and electricity), as well as better access to primary healthcare. Maternal and child mortality rates improved, as did life expectancy.

Gupta and Rodary (2017) acknowledge the country's troubled and violent path to independence but avoid characterizing it as a "failed" African state. They point to the country's distinctive regions and provinces, and see it as a "dynamic part of southern Africa; that is, less as an isolated former Portuguese colonial space or a structural component of the South African political economy but rather one with many connectivities" (Gupta & Rodary 2017: 182). One such distinctiveness is uneven regional development, the roots of which go back to Portuguese occupation. Colonial and postcolonial policies gave preferential treatment to the southern provinces, leading to inequalities in income and health. Evidence for such inequality is provided in Table 2.5, although the data are some years old. Poverty is concentrated in northern states. According to the Oxford Poverty and Human Development Initiative (OPHI), the percentage of the population in severe poverty in Maputo province in 2017 was 11, whereas in Cabo Delgado and Zambezia it was 65 (OPHI 2017).

Although Cabo Delgado is one of the country's most marginalized provinces, it is rich in natural resources, particularly with the discovery of precious gemstones (discussed further in Chapter 6). But profits from Mozambique's extractive industries (including natural gas fields) have flowed to the country's political elites and foreign investors, leaving locals impoverished. Coastal fishing and agricultural settlements have been displaced, fuelling resentment. Cyclones in 2019 and 2020 caused further devastation of coastal communities, destroying crops and fishing villages. Mugabe *et al.* (2021) give further

Table 2.5 Child mortality by province in Mozambique, 2003

Province	Child mortality (per 1,000 live births)
Cabo Delgado	241
Nampula	220
Niassa	206
Tete	206
Sofala	205
Manica	184
Gaza	156
Inhambane	149
Zambezia	123
Maputo	108

Source: Macassa *et al.* (2012).

evidence that Cabo Delgado province hosts tens of thousands of vulnerable households, both as a result of Cyclones Idai and Kenneth in 2019 and the substantial internal displacement of people following incursions by groups linked to Islamic State. Even before the cyclones struck the country, the provision of clean drinking water and sanitation was poor (and spatially uneven), leading to the spread of infectious disease. There are annual outbreaks of cholera during the wet season, particularly in the northern and central provinces. Malaria, the fourth leading cause of death in 2019, is endemic. The spread of Covid-19 has been enabled by overcrowded conditions, especially in Pemba, the capital of Cabo Delgado province. By June 2021 only one Mozambican in 100 had received their first dose of Covid-19 vaccine.

Democratic Republic of the Congo

The DRC in central Africa (Figure 2.8) occupies land area equivalent to that of mainland western Europe (2.3 million km^2). The population in 2021 was estimated to be 93 million, about double that recorded just 20 years earlier.

Its history of colonial rule has been told in graphic terms by Hochschild (2019), who speaks in detail about the atrocities carried out during the "ownership" by King Leopold II of Belgium, who, between 1885 and 1908, treated the region as his personal fiefdom before transferring it to the Belgian state.

The Republic of the Congo secured independence from Belgium in 1960 and elections led to the installation of Patrice Lumumba as prime minister. Lumumba sought to sever ties with western Europe, courting the (then) Soviet Union; as a result, President Eisenhower authorized his assassination (Hochschild 2019: 302), and the local enabler of his death, Mobutu Sese Seko, was later installed as president, acting as dictator for 30 years (and renaming the country Zaire) before being overthrown in 1997. As Hochschild argues, in plundering his country's assets he was little different from the Belgian king.

In her introduction to Hochschild's book, Barbara Kingsolver refers to both the "covert intervention of foreign powers to undermine the newly independent democracy", as well as "a wrestling match of multinational corporations reaching into Congo's deep mineral pockets in a postcolonial free-for-all" (Hochschild 2019: viii). She continues: "Congo is still a rich land of very poor people." Despite such resource riches (see Chapter 6), nearly two-thirds of the population live on less than $1 per day, only a quarter have access to clean drinking water and less than 20 per cent of the population has completed secondary school education.

Since Mobutu's rule ended in 1997, several wars have had devastating impacts on the economy and population, particularly in the east of the country, which borders Rwanda, where there has been mass population displacement and food shortages – and hence considerable increases in mortality. Although life expectancy has risen since 1990 (when it was 52 and 56 years for men and women respectively), in 2017 it was estimated only as

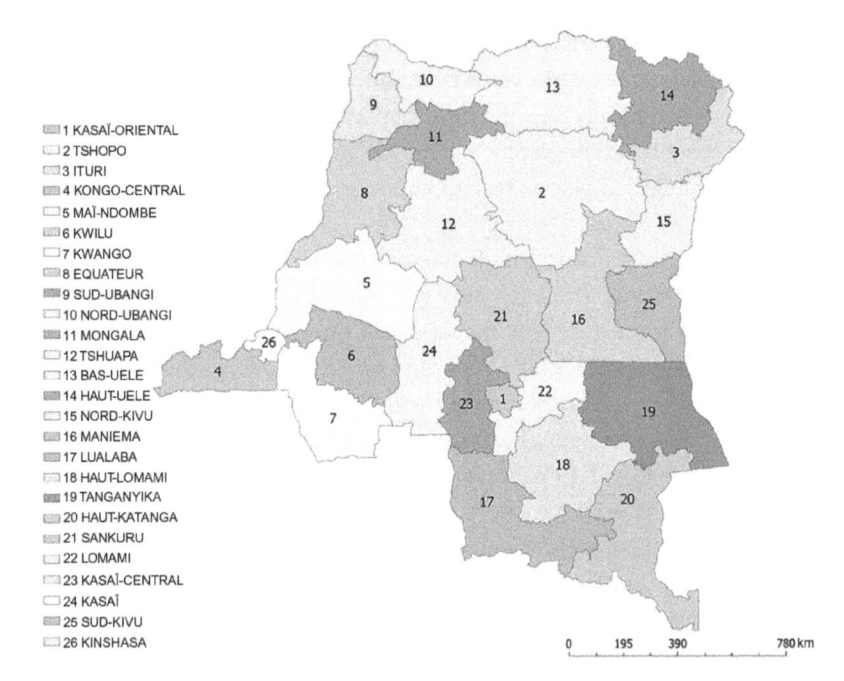

UNEQUAL HEALTH I: DETERMINANTS, REGIONAL EXAMPLES

1 KASAÏ-ORIENTAL
2 TSHOPO
3 ITURI
4 KONGO-CENTRAL
5 MAÏ-NDOMBE
6 KWILU
7 KWANGO
8 EQUATEUR
9 SUD-UBANGI
10 NORD-UBANGI
11 MONGALA
12 TSHUAPA
13 BAS-UELE
14 HAUT-UELE
15 NORD-KIVU
16 MANIEMA
17 LUALABA
18 HAUT-LOMAMI
19 TANGANYIKA
20 HAUT-KATANGA
21 SANKURU
22 LOMAMI
23 KASAÏ-CENTRAL
24 KASAÏ
25 SUD-KIVU
26 KINSHASA

Figure 2.8 Provinces of the Democratic Republic of the Congo
Source: Mappr (n.d.).

60 and 64 years. The leading causes of death in 2019 were malaria and tuberculosis, but evidence suggests that non-communicable diseases such as stroke and ischaemic heart disease are increasing rapidly, as elsewhere in the Global South (IHME n.d.). There has been considerable improvement in levels of child mortality since 2000, although considerable geographical variation remains, with relatively high rates in the southeast (Tanganyika and Haut-Katanga provinces; see Figure 2.8 for a reference map). Malnutrition is the main driver of these rates.

Aside from the indirect effects of conflict in perpetuating malnutrition, there are direct effects on people living in the conflict zones. Such effects are felt particularly strongly by women who are subjected to sexual violence – as explored in the following chapter.

Concluding remarks

This chapter has considered the determinants (social, political and commercial) of health inequalities in a global context, and painted a picture of geographical variations in health outcomes, with a particular focus on child and maternal mortality. As well as examining patterns of variation in five selected countries, it has been emphasized that there are considerable within-country variations. As Farmer *et al.* (2013: 186) have noted, "[S]tudies need locally-specific information as disease prevalence varies within countries ... [R]ecognizing geographic differences in prevalence rates is essential to any health delivery strategy." This is as true of countries in the Global South as it is in the Global North.

Income inequality is a feature of all economies, evidence suggesting that, in most countries, it is increasing. Evidence also continues to mount that inequality is bad for health, although it must be true that absolute poverty dictates health outcomes in many countries in the Global South. As noted earlier, laudable though the Sustainable Development Goals are, the targets are, in the main, hopelessly unrealistic, and tensions remain between striving for better health and the relentless drive for growth – given new impetus following the Covid-19 pandemic.

Only a handful of countries have been examined in some detail here, and I encourage readers to explore others in more detail. The following chapter turns attention away from vignettes of particular countries to examine a number of key themes relevant to the geography of health inequalities, particularly in the Global South.

Further reading

I strongly recommend Boyle (2021: ch. 7) for an excellent introduction to development theory and global inequalities.

There is a vast literature on the geography of health inequalities in the Global North. A good starting point is Bambra (2016), but for a North American perspective see the paper on "eight Americas" by Murray *et al.* (2006). For a popular account of Murray's work in creating the Global Burden of Disease project, see Smith (2015). For a critical examination of the use of quantitative data in global health, see Adams (2016). On links between sustainability and health geography, with a particular focus on gender, see Williams and Luginaah (2022).

Dorling (2013) brings together many of his earlier papers on health inequalities, mainly, but not exclusively, relating to Britain, and is a rich source of typically thoughtful and provocative material. For an excellent broader discussion of health inequalities, I recommend strongly the paper by Arcaya, Arcaya and Subramanian (2015). On links between income inequality and ill health, the classic work is by Wilkinson and Pickett (2010).

In order to understand what progress is being made towards Sustainable Development Goal 3.2 (ending preventable child deaths by 2030), you should see Burstein *et al.* (2019) and their detailed subnational rates of child mortality between 2000 and 2017 and, crucially, the annual reports provided by the Gates Foundation (2022).

References in the text should be consulted for more detail on health inequalities in the Global South. For example, the *International Journal for Equity in Health* (vol. 15, 2016) had a collection of essays on health inequalities in Brazil.

One world region that I have not touched on is the Arab world, and I recommend the collection edited by Jabbour *et al.* (2012).

3
UNEQUAL HEALTH II: KEY THEMES

> We must link ethnography to political economy and ask how large-scale social forces become manifest in the morbidity of unequally positioned individuals in increasingly interconnected populations.
>
> Paul Farmer, quoted in Ecks and Harper (2013: 252)

It is easy to imagine countries within the Global South as "hearths" of infectious disease, an imaginary that is enabled by graphic stories of Ebola in parts of west Africa. This chapter moves away from such tropes, by focusing initially on non-communicable diseases and the role played in these by markets for alcohol and tobacco. I then turn to matters of gender inequality, revealing how this too shapes health outcomes for women in the Global South.

There is a welcome, and growing, interest in inequality among Indigenous peoples, whether in the Global North or Global South, and this is considered next. The impact of the maltreatment of such peoples and the resulting trauma are felt sharply in particular places, and this trauma transmitted from generation to generation.

The previous chapter painted some broad-brush canvases of health in five countries but said little about the health of people living in particular places. The present chapter addresses some of the inequalities in urban areas. It then considers inequalities in mental health, with a particular focus on children, an area in which – alongside social determinants – exposure to violence is of major concern. Links between place and wellbeing have attracted considerable interest in recent years (witness the new journal *Wellbeing, Space and Society*), and the interest extends to places in the Global South.

There is a generally accepted view among those researching health inequalities in the Global North that access to health services plays only a minor role in determining these. Established technologies such as geographic information systems (GISs) can help monitor unequal access, while newer ones based on the use of mobile phones can also help. But the same "inverse care law" (those most in need get the poorest access) applies even in the Global South.

Geographies of inequalities in non-communicable diseases

As noted in Chapter 2, Sustainable Development Goal 3 seeks to "[e]nsure healthy lives and promote well-being for all at all ages"; within this, Target 4 aims to "reduce by, one-third, premature mortality from non-communicable diseases through prevention and treatment". Non-communicable diseases include cancer, cardiovascular disease (heart disease and stroke), lung disease and diabetes, evidence suggesting that almost 80 per cent of deaths from NCDs occur in low- and middle-income countries (WHO 2022a). Epidemiologists recognize four major "proximate" risk factors for such conditions, including smoking, alcohol consumption, poor diet, and lack of exercise, although many such chronic conditions are shaped by "distal" or structural factors beyond individual control or agency (Glasgow & Schrecker 2016; Yang, Mamudu & John 2018). Glasgow and Schrecker (2016) alert us to commercial determinants (Box 2.1). None of us, least of all those living in LMICs, can control the marketing campaigns of the fast food industry or tobacco companies, where film stars may be used to promote smoking (Figure 3.1).

The WHO's recent global report on the state of NCDs refers to "root causes" (WHO 2020: vi). Specifically, it calls for governments to control demand for tobacco and alcohol, and to do so it recommends they raise prices and "enact and *enforce*" (WHO 2020: 2–3, emphasis added) bans on tobacco and alcohol advertising. For each country, progress is monitored in terms of the number and percentage of deaths from NCDs, the probability of premature mortality attributable to NCDs and whether measures to address alcohol, tobacco, diet and physical activity are being achieved. In most low- and middle-income countries such measures show either no, or only partial, progress. Unsurprisingly, therefore, Martinez *et al.* (2020) report that, in 2017, high levels of premature preventable mortality from NCDs were found in countries in southeast Asia, the eastern Mediterranean and Africa. They highlight ten countries (with a population of more than 2 million people) that have the highest levels of premature preventable mortality from NCDs: Afghanistan, the Central African Republic, Uzbekistan, Haiti, Mongolia, Turkmenistan, Pakistan, Ukraine, Laos and Egypt. As a result, the authors call for "more robust, politically committed responses to the social determinants of NCDs" (Martinez *et al.* 2020: e522).

The World Bank monitors data on smoking and alcohol use as part of its Sustainable Development Goals agenda, and some data are presented in Tables 3.1 and 3.2. Those countries (top ten) are shown that had the highest rates in 2018, with comparator data for either 2000 (alcohol) or smoking (2007). They are ranked in order, according to the 2018 rates.

Table 3.1 shows that countries in southeast Asia and the Pacific islands have high rates of smoking, although these have generally fallen since 2007; however, rates of males smoking in Indonesia and Lesotho have increased during the 11-year period. The growth of the e-cigarette market provides a further opportunity for transnational companies to shape consumer preferences (Box 3.1). From Table 3.2 we see that consumption of alcohol has increased since 2000 in some countries of the Global South (Seychelles and Burkina Faso), although high rates of alcohol consumption are not confined to such countries.

Figure 3.1 Indian actor Jackie Shroff promoting cigarettes
Source: https://pbs.twimg.com/media/DxbKezLU8AEAnPt.jpg.

Table 3.1 Percentage of male adults smoking, from SDG Target 3.a, 2007 & 2018

Country	2007	2018
Indonesia	67.9	70.5
Myanmar	77.9	70.2
Kiribati	76.0	68.6
Tuvalu	77.4	66.0
Timor-Leste	82.6	65.8
Bangladesh	63.0	60.6
Laos	75.3	60.1
Solomon Islands	57.2	55.9
Lesotho	42.5	54.7
Georgia	59.0	43.2

Source: https://datatopics.worldbank.org/sdgs/index.html.

Reubi, Herrick and Brown (2016: 181) are sceptical of the "ever more sophisticated epidemiological investigations" of NCDs. They reveal how NCDs came to be seen as a "development problem" (for example, by the WHO and World Bank) for countries in the

Table 3.2 Total alcohol consumption by persons over 15 years (in litres, per capita), from SDG Target 3.5, 2000 & 2018

Country	2000	2018
Seychelles	7.62	20.50
Uganda	13.61	15.09
Czechia	15.00	14.95
Lithuania	13.36	13.22
Luxembourg	14.32	12.94
Ireland	15.13	12.88
Spain	12.43	12.72
Latvia	8.76	12.77
Bulgaria	11.80	12.65
Burkina Faso	8.40	12.03

Source: https://datatopics.worldbank.org/sdgs/index.html.

BOX 3.1 NEW MARKETS FOR SMOKERS

The WHO introduced, in 2003, a Framework Convention on Tobacco Control as a means of addressing the globalization of the tobacco epidemic. Measures proposed include price and tax controls, along with improved packaging, advertising and marketing. As a result, tobacco companies have sought new opportunities to protect profits, one such being developing new products, of which e-cigarettes (vaping) are the most prominent, and potentially lucrative.

The Asia-Pacific region is a major target for e-cigarettes, and the market value increased there from $1.4 billion in 2018 to $2.2 billion just a year later; in Latin America the value was $62.5 billion in 2019. Marketing is undertaken in various ways, including events such as music festivals and the use of high-profile influencers on social media platforms, all of which serve to potentially attract new, young smokers rather than helping existing adult smokers to quit conventional cigarettes. Different countries regulate, or not, in different ways. India bans e-cigarettes, while Australia offers them only on prescription. In many countries in the Global North they are widely available, with retail outlets in every high street.

Public health officials have mixed views about e-cigarettes, some claiming they help smokers to quit while others take a more cautious approach, suggesting that the long-term health risks are uncertain. For some, e-cigarettes help to reposition the tobacco industry as a responsible business, part of the solution to tobacco-enabled NCDs.

For further reading see the excellent resources provided by the University of Bath's Tobacco Control Research Group (www.bath.ac.uk/research-groups/tobacco-control-research-group).

Global South, and they further point to inequities in care and the structural determinants of NCDs expressed by Glasgow and Schrecker (2016). They go on to discuss the importance of culture – the adoption, for example, of different eating habits that are encouraged by subtle, and not so subtle, advertising and marketing.

How are NCDs shaped by place? One compelling example comes from interviews undertaken in a poor part of Cape Town, South Africa, called Khayelitsha (Smit *et al.* 2016). Here, the difficulties of getting access to fresh food, as a result of travel distances or the absence of facilities for growing their own, mean that eating healthily is unrealistic. Equally, sports and playground facilities are badly maintained and unsafe, limiting opportunities for physical activity despite interviewees knowing very well that exercise is important. Such findings parallel those from the considerable volume of research on the geography of health in the Global North, on food deserts, obesogenic environments and the walkability of streets.

Battersby (2017) flags growing interest in obesity as a global health issue but argues that there is still an overemphasis on under-nutrition or malnutrition, and that it is simplistic to regard the latter as merely a Global South issue and obesity simply as a Global North problem. The Sustainable Development Goals fail to acknowledge the rapid nutrition transition; there is no specific target or indicator that addresses obesity, perhaps because of the "powerful voices of Big Food protecting their business interests" (Battersby 2017: 200). Like Smit and colleagues (2016), she turns her focus to South Africa, since it has the double burden of undernutrition for some and obesity for others. Low-income households are most exposed to high-energy, nutritionally deficient foods, as are wealthier households, but the latter can afford better-quality food. The prices of fruit and vegetables have risen more rapidly than those of processed food. Battersby shows how food vendors target children, locating small outlets near schools and around transport hubs. This mirrors what has happened in parts of the Global North.

Non-communicable diseases are leading causes of death in China (Zhou *et al.* 2019). Looking at years of life lost, there are clear regional differences (Table 3.3), with both stroke and heart disease showing significantly higher rates as one moves from east to west (see Figure 2.5 for a map of Chinese provinces).

Table 3.3 Age-standardized years of life lost per 100,000 in China, 2017

Region	Province	Stroke	Ischaemic heart disease
West	Xinjiang	2,300	2,400
	Qinghai	2,700	2,000
Centre	Sichuan	1,800	1,000
	Hubei	2,300	1,600
East	Guangdong	1,300	1,300
	Beijing	940	1,000
China		2,000	1,600

Source: Zhou *et al.* (2019).

Herrick and Reubi (2021) are among several authors to highlight the associations between NCDs and Covid-19. As they put it: "NCDs themselves emerge as a risk factor for infectious disease. These epidemiological interactions and their clear link to environmental and climate change offer fascinating points of future leverage for the NCD community" (Herrick & Reubi 2021: 2). I return in later chapters to both infectious disease and climate change.

Geographies of gender inequality

Sustainable Development Goal 5 seeks to promote gender equality and remove discrimination. To help assess this, a Gender Inequality Index has been devised (Box 3.2), which has been used quite widely to examine associations with health outcomes.

In the previous chapter we looked at some evidence on child mortality, but there is more to be said about sex differences in such mortality, and the determinants thereof.

Iqbal *et al.* (2018) look at the association between the GII and sex-specific child mortality. They find that the more gender unequal a society is, particularly in low- and

---

BOX 3.2 THE GENDER INEQUALITY INDEX

The Gender Inequality Index (GII) was devised in 2010 by the United Nations Development Programme (UNDP) in its *Human Development Report*. It combines data on three dimensions: health; empowerment; and the labour market. The first of these is measured by the maternal mortality ratio and adolescent birth rates; the second by the proportion of parliamentary seats occupied by females and the proportion of adult females with at least some secondary education; and the third by the labour force participation rate of females and males. These data are combined and standardized to give an index that varies from 0 to 1; the higher the value, the greater is the gender inequality in a given country.

For example, in 2019 the GII in all Scandinavian countries was less than 0.05. In China the GII was 0.168, in Brazil it was 0.408 and in India it was 0.488. In Mozambique it was 0.523, and in the DRC 0.617. The most gender-unequal country was Yemen (0.795). In Afghanistan the index in 2019 was 0.655, a figure that is likely soon to be distressingly out of date given the Taliban regime's treatment of women and girls.

Not all scholars agree that the GII is helpful. In a wide-ranging and technical appraisal, Permanyer (2013) suggests that its use of women-specific and "men versus women" indicators is unhelpful, while its use of maternal mortality data ignores the fact that maternal mortality is less to do with gender and more with poverty. She would prefer a "Women's Disadvantage Index".

For further details, including data, see (UNDP n.d.).

---

middle-income countries, the less likely are girls to survive. Some attribute this to biology, but more plausible is that young girls' lives are valued less; they may have less access to health-promoting resources such as immunization. In India, for example, it has been shown (Corsi et al. 2009) that girls are significantly less likely to be vaccinated than boys. This varies between states: those with the lowest girl-to-boy ratios of immunization coverage are Punjab, Haryana and Bihar, in the north of the country (see Figure 2.4); there, 10 to 14 per cent fewer girls than boys are fully immunized, despite overall immunization rates in those states being above average in the country as a whole.

This is a good example of gender-based structural violence (Box 1.1); but there are others. The availability of antenatal screening and determination of sex by ultrasound means that selective abortion is an option in India, where sons are preferred over daughters. Across the globe the sex ratio at birth is approximately 94 girls per 100 boys. In India, in 2011 the ratio was 89.9 (Kim et al. 2019). There is wide variation in the sex ratio across states and districts, and between villages within districts. The same is true of the child sex ratio (proportion of girls and boys aged up to six years). This ranges from 80 to 82 girls per 100 boys in Punjab and Haryana states (northwest India) to 96 to 97 girls per 100 boys in Jharkhand (east India) and Kerala in the south. Kim and colleagues (2019) provide very detailed maps. In 2014 the national and state governments introduced schemes called *Beti bachao, beti padhao* ("Save daughters, educate daughters"), to rebalance sex ratios at the local level. This may go some way to redressing what Amartya Sen has referred to as "India's missing millions" (Sen 1990), or "natality inequality as opposed to mortality inequality" (Gatrell & Elliott 2015: 135–6).

As noted in Box 3.2, maternal mortality is one component of the GII. The group Physicians for Human Rights (2002) have undertaken research on maternal mortality in Herat province, Afghanistan, where women have a high risk of death during pregnancy (593 maternal deaths per 100,000). This research was based on a survey of 4,500 women and interviews with both women and their carers in 2002. Only 11 per cent reported having received any prenatal care, and only one of the 13 districts in Herat province had obstetric care facilities meeting WHO standards. Midwife Tamar said:

> Most babies are delivered in the villages by untrained birth attendants. If there is a very sick woman or a complicated pregnancy, we refer them to [the city of] Herat but I do not know how many can go since it is so far from here … most of the women I see are malnourished … There is no sanitation in the village … women die at home or on a donkey on the way to Herat.

Physicians for Human Rights recommended that the Afghan government make the reduction of maternal mortality a national priority. Whether regime change under the Taliban in 2021 endorses this seems highly unlikely. Gender discrimination means that an adequate female health workforce becomes difficult, if not impossible, to sustain, threatening population health across the country. Access to health services is considered later in this chapter.

Reichenheim *et al.* (2011) document the importance of physical, not structural, violence against women, including domestic violence, as a public health problem in Brazil (see also Heimer *et al.* 2022). There is a strong regional pattern to this, with higher prevalence in the north of the country, where there is both a stronger patriarchal culture and more gender inequality compared with the more developed south (Reichenheim *et al.* 2011). Twenty per cent of women in the North and Northeast reported an episode of physical force, compared with 12 per cent in the Central-West, Southeast and South. In 2008 women accounted for over 27,000 of the 31,000 cases of grievous bodily harm registered at police stations in Rio de Janeiro; in more than half these cases the perpetrators were either present or former partners of the victims.

Acknowledging the considerable difficulties in establishing the extent of sexual violence in the Democratic Republic of the Congo, Johnson *et al.* (2010) undertook fieldwork in three eastern provinces (Nord Kivu, Sud Kivu and Ituri; see Figure 2.8 for reference) during 2010. Their findings indicate widespread sexual violence and human rights abuses in the previous decade; 40 per cent of women and 24 per cent of men in their sample of about 1,000 reported sexual violence linked to conflict. Extrapolating from that sample, the authors suggest that maybe 1.31 million women and three-quarters of a million men are survivors of sexual violence and therefore in need of specific forms of violence-related care; however, the prospects of receiving such are remote. Improving women's health demands a recognition of both the structural and physical violence that they suffer. These affect their access to healthcare, especially if this is dictated by childcare and other domestic demands, as well as their work (often agricultural) outside the home. As Farmer *et al.* (2013: 309) put it, "Gender inequity and poverty enjoy a noxious synergy."

Child marriage is another dimension of gender inequality. Patton *et al.* (2016) report that, globally and annually, 15 million girls marry before the age of 18 years, and that in countries of the Global South (particularly in sub-Saharan Africa) one-third of girls are married before 18 years; one in nine before the age of 15. As a result, they may have fewer opportunities to complete school and face early pregnancy, higher rates of maternal and infant mortality, and a higher risk of HIV/AIDS and other sexually transmitted diseases.

Geographies of Indigenous health inequalities

In writing about Indigenous health there are many examples upon which to draw. Here, I reflect briefly on the health of those colonized in North America: Native Americans in the United States and First Nations groups in Canada.

Table 3.4 Age-adjusted percentages of Native American adults in fair or poor health, by area type, 2014–18

Area type	AIAN adults (> 18 years)	US adults
Large MSAs	19.0	11.1
Medium/small MSAs	22.8	12.4
Rural areas	20.7	15.6
All areas	20.6	12.1

Note: MSA = metropolitan statistical area.
Source: Villaroel *et al.* (2020).

There are statistically significant differences between the health of the AIAN (American Indian and Alaska Native) populations and those of other US adults; this is true whether they are living in urban or rural areas (Table 3.4). One-fifth of Native Americans report fair or poor health, compared with only one in eight adults in the wider population.

Other quantitative work by Findling *et al.* (2019) reports evidence of inequalities among Native Americans compared with whites, in terms of pay, access to healthcare, interactions with police, violence and threats towards individuals and family members. Those living on tribal lands, or areas with high concentrations of Native Americans, appear to suffer more discrimination.

History matters. "AIAN people have been devastated by disease, warfare, forced migration, cultural genocide, racism, and poverty. Likewise, cultural practices, including languages, educational systems, spirituality, and the daily practices of everyday life were systematically attacked, oppressed or outlawed" (Walters *et al.* 2011: 181). Such historical events and trauma are embodied and transmitted across generations; the determinants of health are historical as well as social.

Listening to the voices of Native Americans enriches our understanding of the discrimination they incur, and the impacts on health. Giordano *et al.* (2009) have collected rich data from Native Americans living in California, using focus groups and interviews to reveal concerns about loss of cultural roots, violence, drug and alcohol abuse, depression and suicide. Historical trauma is a theme that recurs (Box 3.3). As one community leader puts it:

> If you treat the symptom of alcoholism and not have it in the context of historical trauma, you are to miss what really the whole family is suffering from – what the whole generation is suffering from. So, this is my big concern: that historical trauma doesn't really get assessed as a diagnosis, it doesn't get treated.

Another says: "I am dealing with people who have been disenfranchised, and their mental illness originates in the system around them, the environment, the surrounding historical trauma. They are not crazy; they are people responding to trauma in their life."

BOX 3.3 HISTORICAL TRAUMA

If population groups have, historically, been subjected to colonialism, slavery, war or genocide, the ill health or disease that results from such trauma may manifest itself not only in those directly affected at the time but also among those in subsequent generations. Rather than a single violent act, traumatic events may endure over a long period of time and affect many Indigenous people (Sotero 2006).

Historical trauma may therefore be seen as a structural determinant of the disparities or inequities in health incurred by minority ethnic groups, such as African Americans absorbing the longer-term impacts of slavery, or Indigenous populations in Canada, Australia or New Zealand subjugated by what became a majority population. Such subjugation takes the form of physical violence, displacement, economic domination and environmental and cultural dispossession (Sotero 2006). The longer-term consequences of these insults are felt – psychologically, and even physiologically – across generations. The stories that are told and passed on from one generation to another get absorbed or embodied, with consequences for individual and population health.

Among many examples of historical trauma, consider recent evidence in Canada of First Nations children being removed from their homes to boarding schools, the aim being to "educate" them out of their native cultural traditions and into new ways of life. One survivor, Jack Kruger, recounts that he was beaten for using his native Sylix language, nsyilxcən. "I lost my language for 40 years because they told me it was the 'devil's tongue'. I was brainwashed so badly. I didn't speak to no one in it – and I didn't let anyone speak the language to me." (Voce, Cecco & Michael 2021). First Nations communities are now unearthing hundreds of graves of children forced to attend such schools, the children dying as a result of abuse, starvation and neglect. The trauma of such appalling treatment is felt not only by the families of those affected but by the wider native communities.

For further details and the relationship of historical trauma to public health, see Sotero (2006).

Brown *et al.* (2012) draw on the concept of structural violence in examining health inequity among Indigenous Canadians, for whom the establishment of reserves that offer few resources, along with the removal of Indigenous children to residential schools and foster homes, have had devastating impacts over many generations. Again, this is historical trauma.

There is a wealth of evidence that reveals inequalities in health between Indigenous and settler Canadians. Data from the Public Health Agency of Canada shows that life expectancy for First Nations men and women is about six years less than for non-Indigenous Canadians, while for the Inuit population the difference is about 15 and ten years for men and women respectively. The same inequalities manifest themselves in access to healthcare. Burnett *et al.* (2020) document problems of accessing care in northern Ontario, where First Nations people living on reserves face the challenges of getting to health centres when roads are closed in winter. When seen by health professionals, some respondents reported being treated roughly and disrespectfully, while the absence of any continuity of care was problematic. Echoing our earlier consideration of decolonizing global health, the authors assert that, although we need to acknowledge Indigenous perspectives on health and healthcare, "this must be done alongside systemic change and a broader decolonization process that returns stolen land and resources and supports Indigenous sovereignty and self-determination" (Burnett *et al.* 2020: 8).

Despite this body of evidence on such health inequalities, whether of outcomes or care, there is an inherent danger of stigmatizing the places where Indigenous people have lived for hundreds of years. In a powerful paper written about communities in northern British Columbia (BC), Aldred *et al.* (2021) argue that "the more stories about underserviced, ruined, and resource-extracted lands that are circulated about northern BC, the more people of northern BC are conflated with that place and understood as toxic". Nonetheless, as shown above, Indigenous peoples' health is shaped by both historic and contemporary (physical and structural) violence, demonstrated most clearly in systemic racial discrimination. This statement holds for Indigenous peoples in other countries, such as Australia and New Zealand.

Geographies of urban health inequalities

Sustainable Development Goal 11 is to "make cities and human settlements inclusive, safe, resilient and sustainable" (Table 2.1), and a key focus is on people living in informal settlements or "slums" (taken here to be residential areas with substandard housing, insufficient living space and inadequate access to clean water and sanitation; see Lilford *et al.* 2019 for various definitions of "slum"). All these characteristics are determinants of health, and those living there are "triply disadvantaged: they tend to be excluded – spatially, socially and economically – from the opportunities that other city dwellers enjoy" (UN Statistics Division 2017). Overcrowding means that health is impacted by those living in the same or nearby dwellings; there is more competition for space and resources, with infections likely to result from the lack of refuse collection, as well as both poor sanitation and the easy transmissibility of infections (Figure 3.2).

The UN reports that, in 2014, 880 million urban residents lived in slums. In sub-Saharan Africa more than a half (56 per cent) of urban dwellers were reported to live in slum conditions. In central and southern Asia the proportion was 32 per cent, and it was 21 per cent in Latin America and the Caribbean (UN Statistics Division 2017). What are the health impacts of such living conditions? Here, I focus on a small number of settings.

Figure 3.2 Dharavi slum in Mumbai, India
Source: iStock.

Bentley *et al.* (2015) report that half the world's underweight children lived in India in 2012 and that this level was higher in the country's informal settlements than in other urban areas. In such settlements in the cities of Delhi, Lucknow and Mumbai the percentage of children who were stunted (low height for age) was, respectively, 74 per cent, 63 per cent and 47 per cent; according to national data in 2015/16 the percentage for the country as a whole was 38 per cent. Bentley *et al.*'s (2015) detailed study of households (and 7,450 children under five years of age) in 40 of Mumbai's informal settlements revealed that only 5 per cent of children had what was described as a minimally acceptable diet, with sugary and salty snacks common among children under two years of age. The poorest children received more snack foods, suggesting that their low cost makes it an easier option for busy parents.

Vearey *et al.* (2010) note that Johannesburg (population 5.9 million in 2021) has a quarter of its population in informal settlements, fuelled by internal migration. Comparisons of households in one informal settlement with those in other inner city areas reveals that only 36 per cent in the former have running water compared with 83 per cent in the latter. One-quarter do not have access to a toilet, instead making use of the open bush, and, although almost all households in the inner city area have a weekly refuse collection, this is available to only a quarter of those in informal settlements. All these factors are determinants of likely poor health.

Szwarcwald *et al.* (2011) contrast slum census tracts in Rio de Janeiro, Brazil, with others classified as poor, intermediate and rich (all based on mean monthly income). Disability scores show a clear gradient from slum to rich areas (Table 3.5), and significantly higher scores for women.

The authors also draw attention to other dimensions of slum living in Rio, indicating that poor young people are drawn into criminal activity. Increasing conflict between rich and poor and the lack of suitable employment make it tempting for teenagers and young adults to earn money and join gangs; disputes over control of drug sales, involvement in robberies and kidnappings lead, the authors argue, to early and violent death. The contrasts between built environment and living conditions in other cities, such as São Paulo, are often stark, as iconic photographs illustrate (Figure 3.3).

Table 3.5 Disability scores in contrasting census areas in Rio de Janeiro, 2006

	Slum	Poor	Intermediate	Rich
Women (18–59 years)	1.31	1.17	1.06	0.37
Men (18–59 years)	0.87	0.37	0.35	0.33

Note: Scores are based on a five-point scale in response to six questions asking about ability to perform various daily tasks; higher scores reflect less ability to perform such tasks.
Source: Szwarcwald *et al.* (2011: 520).

Figure 3.3 Contrasting places: the Paraisópolis favela and luxury buildings in São Paulo, Brazil
Source: iStock.

I noted above the prevalence of violence against women in Brazil, and this is a public health problem in Indian cities too, where the National Family Health Survey in 2015 reported that 31 per cent of women of reproductive age had experienced violence. Research by the UN's Safe Cities Free from Violence against Women and Girls programme found that, in Delhi, 92 per cent of women up to 50 years of age had reported experience of various forms of sexual harassment or violence in public spaces (Zietz & Das 2018). Among men, the lack of space and permanent employment, poverty and alcohol abuse are all factors generating stress, which spills over into violence against women in the same households. A systematic review of violence against women in Indian slums (Jungari et al. 2022) suggests that the prevalence of physical violence ranged from 9 per cent to 63 per cent, psychological violence from 11 per cent to 86 per cent and sexual violence between 2 per cent and 59 per cent; evidently, violence against women in many Indian slums is widespread.

Zietz and Das (2017) conducted fieldwork in two slum areas of Mumbai, close to a large dumping ground for refuse where many residents work collecting, sorting and recycling garbage. Focus groups comprising adolescent and young men were conducted. Some reported that they justified harassment of women on the basis that only "bad girls" are harassed in public. As one man in the 18- to 24-year-old range put it: "First, we look at the girl. If she is of good nature, then we don't do anything. If we feel that she is of loose morals, then we start teasing her." Another suggested that "if any girl argues with these boys, they will beat her. Girls change their walking routes because of teasing." Being a gang member, or a user of drugs and alcohol, were seen as risk factors for the perpetration of sexual harassment, which was normalized as part of daily life.

Those living in slums are invariably stigmatized. They are othered – excluded spatially, economically and politically, all of which exclusions and stigmatization are determinants of poor health. But some authorities seek to improve living conditions by upgrading the housing and infrastructure. Slum upgrading ranges from small-scale projects to introduce better water supplies, street lights and paved roads to much larger-scale interventions that involve both physical infrastructure and social programmes (Corburn & Sverdlik 2017). Such upgrading has clear health benefits.

This section is called "urban health inequalities" but the focus has been exclusively on those living at the very margins (often spatially, always socially) of cities. There is much more that could be said about urban living in the Global South and how this affects health, as well as urban–rural contrasts. But my focus on the most "othered" of urban populations offers a link to later chapters, since people who live in slums are particularly vulnerable to infectious disease and to the effects of global warming. Poorly built dwellings are hardly equipped to deal with storms and tropical cyclones, while inadequate sanitation and water supplies are perfect conditions for the spread of disease. The health impacts of these are considered later. For now, it should be noted that, although informal settlements and their residents are invariably stigmatized, evidence from places such as Dharavi (in Mumbai) suggests considerable success in managing Covid-19 (Kumar et al. 2020). Other research from Brazil (Gillam & Charles 2019) suggests that community leaders in such settlements can help improve the lives and wellbeing of those living there, countering the impact of inequality and racism by helping to introduce new community-based programmes and facilities. It is important not to cast all those living in such places as victims; for example, detailed research in Dhaka, Bangladesh, reveals how slum residents come together to counter dominant discourses of public health risk (Fattah & Walters 2020). Similarly, Amin and Richaud (2020), in their research on migrants to Shanghai, China, find that stress was moderated by lived experiences and encounters in ordinary settings such as libraries, cafés and bookshops.

Geographies of mental health inequalities

The very ambitious SDG 3 seeks to "[e]nsure healthy lives and promote well-being for all at all ages" (Table 2.1 above). Target 3.4 aims to "reduce by one third premature mortality from non-communicable diseases through prevention and treatment *and promote mental health and well-being*" (emphasis added).

Lovell, Read and Lang (2019) suggest that the Global Burden of Disease programme has given prominence to global mental health; indeed, the GBD remains a good source of data on mental health. It suggested that in 2017 the number of people living with any kind of mental disorder was 792 million – about one in ten of the world's population. Of that number, 284 million had anxiety and 264 million had depression (Our World in Data provides, for all countries, excellent interactive visualizations, including graphics on change over time and the proportion of disease burden attributable to specific conditions). For example, Figure 3.4 shows age-standardized disability-adjusted

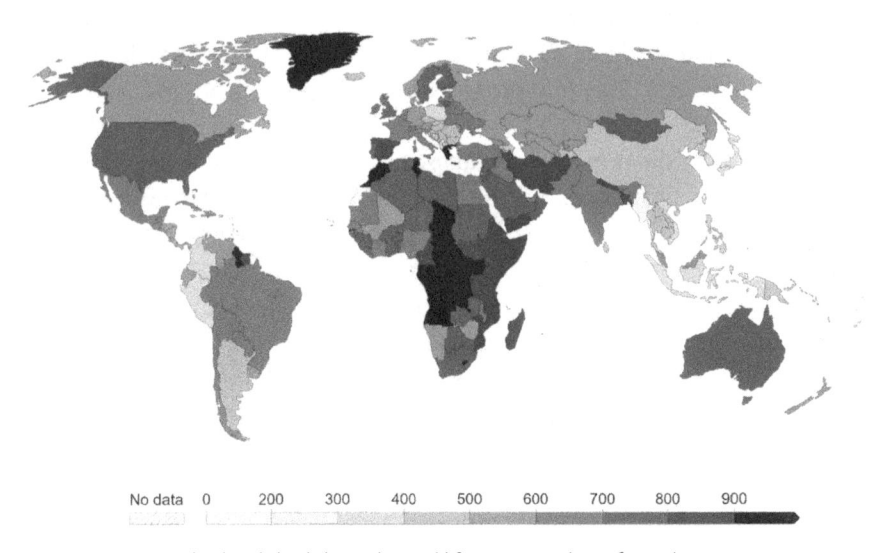

Figure 3.4 Age-standardized disability-adjusted life years resulting from depression
Source: https://ourworldindata.org/grapher/dalys-depression-age-std-rate.

life years (DALYs) resulting from depression. Drawing from this, note that among the very highest rates are Lesotho (994), Iran (885), Palestine (803) and the Central African Republic (729), although in the Global North rates for the United States and Australia are also high: 737 and 719 respectively.

Patel *et al.* (2018a), writing on behalf of the *Lancet* Commission on Global Mental Health, and reviewing in detail the history of global mental health initiatives, suggest that global mental health rests on four "foundational pillars". The first of these is that mental health is a global public good and is of relevance to sustainable development in every country. Second, poor mental health ranges from issues that might be mild and time-limited to very severe, chronic conditions. Third, good mental health is a fundamental human right. Last, mental health is significantly related to social and environmental factors.

On the last of these, the social determinants of poor mental health include poverty, income inequality, family and community violence, environmental quality, environmental disasters and armed conflict leading to forced migration, all of which are place-based or have a spatial context that is both local and often spills across international boundaries. These determinants of course intersect. For instance, young people who have been displaced by war can find themselves living in poverty and at serious risk of mental illness, while other young people who are out of work may use alcohol or drugs, which also place them at risk of mental ill health.

A clear focus on structural determinants of mental ill health is at odds with a medical (and specifically psychiatric) model of illness that promotes the use of drug therapies (see Box 3.4) rather than the talking therapies now quite commonly used in Western settings. In other words, an uncritical approach to psychiatry situates mental health problems squarely at the individual level, without (as a critical approach would suggest) acknowledging properly the role played by factors well beyond the control of the individual. This tension arises between proponents and opponents of the Movement for Global Mental Health (MGMH: www.globalmentalhealth.org), a coalition that works primarily for those in low- and middle-income countries. For its critics, MGMH sees "distress as symptomatic of 'neuropsychiatric disorders' rather than as responses to socio-politico-economic conditions of conflict, entrenched social inequality, and chronic poverty (to name but a few of the lived realities of global capitalism and liberal individualism)" (Mills & Fernando 2014: 189). In India, Popa (2020) reports research that suggests living in a temple for brief period aids improvements in mental health and wellbeing. Interestingly, this chimes with the plethora of research on therapeutic landscapes (see, for example, Sperling & Decker 2007).

Focusing on one of the key structural determinants of mental health – armed conflict – it is evident that some such conflicts (such as in the Palestinian Occupied Territory: Box 3.5) last for many years, and cut across territorial boundaries.

BOX 3.4 ANTIDEPRESSANTS IN THE GLOBAL SOUTH

Nikolas Rose (cited in Brown *et al*. 2018: 227) suggested that mental ill health provides many opportunities for introducing drug treatments to new markets in the Global South; between 1990 and 2000 this market doubled in value in countries in South America and increased by 137 per cent in Pakistan. Destigmatizing mental health in some countries, Japan in particular, along with promoting drugs at conferences, helped the launch of new antidepressants (selective serotonin reuptake inhibitors, or SSRIs), such as Prozac.

Ecks and Basu (2019) suggest two possible explanations for the rapid worldwide increase in the use of antidepressants. One is that globalization has led to massive urbanization, migration, worsening social cohesion and widening social inequalities; these cause stress and lead to a demand for medication. The other links the increasing use of antidepressants to aggressive marketing by pharmaceutical companies: the drugs create a need. "Diffuse states of sadness, which earlier went without therapeutic intervention, are becoming increasingly medicalized and treated with drugs" (Ecks & Basu 2019: 87). Lovell, Read and Lang (2019) note that these drugs are not only exported from countries in the Global North but are distributed within the Global South by large and small pharmaceutical companies alike (in India, for example). Empirical research in India by Ecks and Basu supports this assertion.

There is another strand to the transfer of drugs produced for the Global North (particularly American) market to other parts of the world. This strand relates to painkillers, the use of which has spawned a so-called opioid crisis in the United States, where some have become addicted to prescription drugs and over 10 million people are reported to have misused prescription opioids in 2019 (www.hhs.gov/opioids/statistics/index.html). Given negative publicity, the market for some painkillers in the United States is declining, leading companies such as Purdue Pharma, marketers of one brand (OxyContin), to seek new business in Brazil and China. Referring to such new markets, the former commissioner of the US Food and Drug Administration warned: "It's right out of the playbook of Big Tobacco. As the United States takes steps to limit sales here, the company goes abroad." Promotional videos by the spin-off company of Purdue Pharma (Mundipharma) boast that "we're only just getting started" (quotes taken from a *Los Angeles Times* investigation: Ryan, Girion & Glover 2016). Whether such a take-off benefits US companies or Indian ones (as suggested by Lovell, Read and Lang and Ecks and Basu) remains to be seen.

For a much broader treatment of the role of pharmaceutical companies and medical research in global health, see Greenhough (2018).

BOX 3.5 CHILD MENTAL HEALTH IN GAZA

There is plenty of evidence concerning the chronic suffering, and exposure to violence, incurred by 1 million children living under military occupation in Palestine, whether in the West Bank or the Gaza Strip. Movements of both Palestinians and material goods are restricted, affecting household income, food supplies and access to healthcare. Almost 60 per cent of Palestinians (and three-quarters of those in the Gaza Strip) live on less than $1.90 a day.

Massad *et al.* (2011) conducted research on pre-school children, assessing the impact on their wellbeing of direct and indirect exposure to violence and material deprivation. Among the 350 children sampled, and mothers' reports on their behalf, the most common exposures to violence were hearing the sounds of jet fighters and the shelling of houses, and seeing pictures of mutilated bodies. Over half the households lacked money to pay bills and on most days of the week the children had no meat, fish, fruits or vegetables to eat. Two-thirds of Palestinian mothers reported poor emotional functioning among children. Curiously, the authors say they "cannot examine causality" and that mothers might be biased in their reporting of children's symptoms; further, and not particularly helpfully, they add that "the public health imperative would be to remove exposures to the stresses of war and violence" (Massad *et al.* 2011: 11).

More than ten years after this study, what has changed? It is left to charitable organizations such as UNICEF to tell the stories that lie behind bland quantitative studies such as that described above. As UNICEF puts it: "Being a child in the Gaza Strip has always been extremely difficult, even before the recent escalation [in May 2021]. For some children, this was the fourth conflict they lived through. No place is safe for children across the Gaza Strip" (UNICEF 2021; see also Figure 3.5). At the time of writing, almost 800,000 people lack access to piped water, given the damage done by shelling to infrastructure, leaving children at risk of contracting waterborne diseases.

The Euro-Mediterranean Human Rights Monitor (an independent, non-profit organization) surveyed over 500 children in Gaza and has reported that, as a result of the May 2021 violence, over 90 per cent of children in Gaza now suffer from some form of conflict-related trauma (Euromedmonitor 2021). Across the border, Israeli children also experience post-traumatic stress disorder symptoms, much more so than those living elsewhere in Israel. Life in Gaza exposes many children to violence that is not only structural but very real.

This is just one of many distressing examples of the mental health burden faced by children in zones of conflict and war. For another set of stories on "invisible wounds", affecting children in Syria, see McDonald *et al.* (2017).

In the previous chapter I reviewed some of the evidence on income inequality and health, but the literature on the association with mental health is inconclusive, certainly in low- and middle-income countries. A good example is South Africa, where Adjaye-Gbewonyo *et al.* (2016) find no evidence of a relationship. However, using data on depressive symptoms for almost 19,000 individuals and income data for 8,000 households located in 53 municipalities, Burns, Tomita and Lund (2017) find that, as income inequality increases and household income decreases, the risk of depressive symptoms increases. As the authors explain:

> If one is poor, one is dealing with the difficulties of poverty and the accompanied higher risk for depression. But if one is poor and living in a region where the differences in material conditions are very significant and apparent, then social comparisons are likely to be distressing as one experiences feelings of failure, poor self-worth and shame. (Burns, Tomita & Lund 2017: 7)

Turning to another determinant of mental health – environmental quality – there is plenty of research evidence to show that access to green space (parks, gardens, wooded areas) is beneficial to the mental health of those living in countries of the Global North. Recently, some authors have sought to establish whether the same relationships exist elsewhere in the world. Building on existing research in Cali, Colombia (a city of 2.4 million), that has highlighted considerable inequalities in access to green spaces,

Figure 3.5 Palestinian children in Gaza Strip
Source: iStock.

Hong et al. (2021) have explored the links between such access and the mental health of 2,400 inhabitants. The specific question asked was: "Thinking about your mental health, which includes stress, depression, and problems with emotions, for how many days during the past 30 days was your mental health not good?" Although their analysis suggested a relationship between proximity to green space and self-reported health, there was no such relationship to mental health. It is questionable whether people living in informal settlements or slums might prioritize access to green spaces over the improvements in sanitation and secure water supplies. Better physical infrastructure would surely confer far greater benefits for both physical and mental health and wellbeing.

Earlier work in Delhi, India, was more suggestive of a relationship. Mukherjee et al. (2017) examined the association between proximity to parks and depression among adults living with chronic conditions such as diabetes, asking if such proximity helps buffer such individuals against poor mental health. As with the Colombia study, the research made use of GISs to define proximity, and statistical analysis to explore the association. Mukherjee's study controlled for probable other effects on depression, such as socio-economic status, and the authors find that the closer people were to larger parks, the lower their levels of depression. The authors link their research to the "smart cities" programme in India, a component of which is to improve urban sustainability through better provision of green spaces.

As Fisher et al. (2021) note, this topic has been under-explored in settings within the Global South. They investigated how people's perceptions of green and blue spaces – as well as natural sounds, such as those from birdsong – impact on wellbeing in Georgetown, Guyana. Taking 57 sites, and measuring environmental characteristics at each, they asked passers-by to rate their "momentary wellbeing". Sites that were considered safe, biodiverse and rich in bird species and birdsong were shown to be restorative and contributing to wellbeing. Yet, as we see below, wellbeing is unequally distributed among populations in the Global South.

Inequalities in wellbeing

Among many definitions, the New Economics Foundation considers wellbeing "as the extent to which an individual or group experiences their life as going well, based on experiencing positive emotions, functioning well and meeting basic psychological needs" (Aked et al. 2008). There is a considerable literature on therapeutic landscapes that considers how natural, as well as built, environments shape healing but also wellbeing (Bell et al. 2018). Since the concept was first introduced over 25 years ago it has expanded to include some examples in the Global South, not least among those living in poverty and zones of conflict.

A good example is research on the Kaqchikel people living in a poor, agricultural highland region of Guatemala: "The Kaqchikel have long-standing experience of discrimination, forced displacement and poverty" (Sperling & Decker 2007: 233). These authors' invitation to take photographs of meaningful places revealed, among other things, that nature provided places of escape, particularly for men used to working in the fields, but also those going to the forest or a nearby lake to reflect on and enjoy their surroundings. In contrast, whereas women found the home important to their wellbeing, it was a place of work and at times a lonely setting. Madge (1998) sees therapeutic landscapes more metaphorically, as sets of particular beliefs about, experiences of and practices of healthcare in places – such as in The Gambia, where she researched such landscapes among the Jola people. Beliefs include seeing illness as having a spiritual or social origin, not only a biological one, and herbal remedies working alongside biomedicine, with the latter more commonly used if health centres were accessible.

Research on wellbeing moves us away from what some see as a deficit view of health, which stresses survival, to a more positive focus on the resources on which people draw to maintain good health. At a global level, the *World Happiness Report 2020* (Helliwell *et al.* 2020) considers the relationship between subjective wellbeing and a Sustainable Development Goal index that measures how well countries are doing in currently meeting SDG targets. Results suggest a non-linear relationship between the two (Figure 3.6).

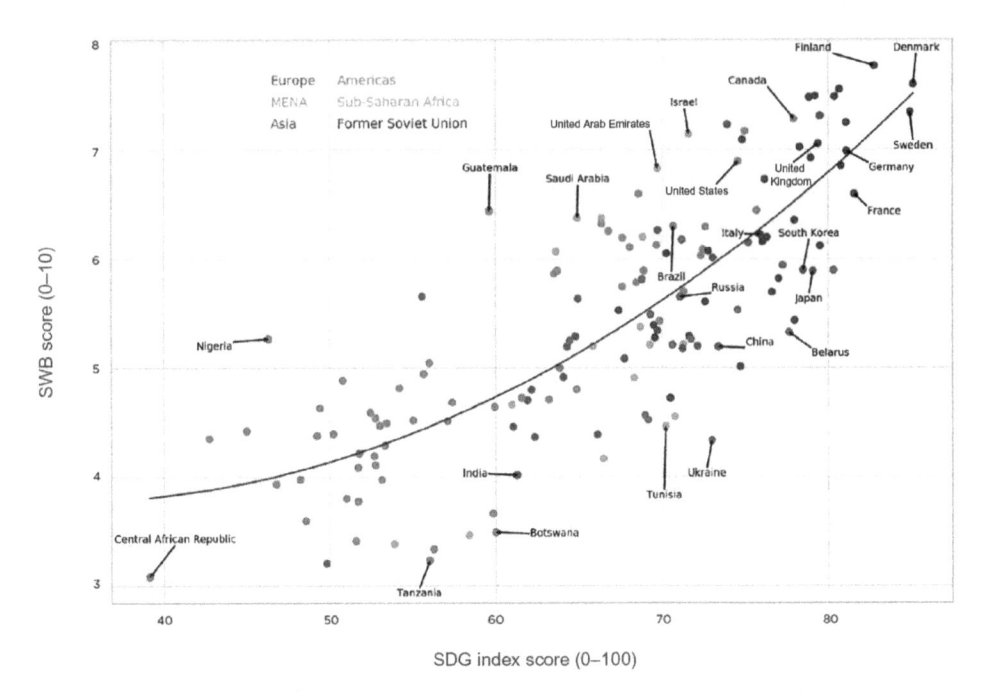

Figure 3.6 Association between sustainable development and subjective wellbeing
Notes: MENA = Middle East and north Africa; SWB = subjective wellbeing.
Source: Helliwell *et al.* (2020).

As is often the case, interest lies more in the outliers. For example, Botswana and Guatemala have identical SDG index scores, but the latter has much higher overall wellbeing; life expectancy is higher in Guatemala, while HIV/AIDS remains the leading cause of death in Botswana.

More revealing than these broad quantitative indicators are qualitative data that describe people's understanding of wellbeing (Camfield, Crivello & Woodhead 2009). A good example comes from research in Ghana. Kangmennaang and Elliott (2019: 178) argue that the current wellbeing literature "ignores socio-cultural, ecological and collective discourses that accompany the 'good life' in other geographical contexts". As they point out, a political ecology approach is needed to show how structural (political, social and economic), as well as ecological, processes shape individual and community wellbeing. The authors interviewed key informants (such as policy-makers) as well as holding focus group discussions with adults living in different communities. Conceptions of wellbeing varied from region to region. For example, in the north of Ghana "community members around you support you and you support them … then there is a social safety net that you can always rely on in times of trouble in the community". By way of contrast, an interviewee in the south referred to being able to "express and exercise … individual creativity and individual desires" (Kangmennaang & Elliott 2019: 182). Most Ghanaians in the sample acknowledged the importance of having sufficient money to meet basic needs, including security of food and water, and good housing. Others were aware of inequality. As one put it: "It's like the rich keep getting richer. You see the difference between the rich and the poor, right now, if we go to the market and I have money, and this gentleman sitting next to me does not, you will see the difference in what he will buy and what I will also buy" (183).

Many of the issues discussed in this chapter can be addressed with good access to services. Improvements in the burden of non-communicable disease and mental health, in the incidence of women and children suffering abuse and in the quality of life for indigenous populations can all be enabled by the better provision of, and access to, health services. What evidence is there of inequalities in such, particularly in the Global South?

Geographies of inequalities in the provision of, and access to, health services

The provision of, and access to, health services tend to be downplayed when considering the determinants of health inequalities in the Global North. Wider determinants are generally recognized to be income, education and working and living environments. But, since large proportions of the world's population have poor access to health services, it is essential to focus attention on inequities not just of health outcomes but of healthcare provision and access too. The inverse care law applies in the Global South as well as in the Global North; good-quality care may be given to those needing it least (Box 3.6), and the attention has to be a geographical one.

In Brazil, the More Doctors programme, which was designed to increase the number of doctors working in more remote areas and which initially saw several thousand doctors come from Cuba, has withered (de Oliveira Andrade 2020), and the Covid-19 pandemic has worsened the situation. One of Brazil's leading health research institutes, FioCruz, suggested in March 2021 that the country was facing its worst ever health crisis (BBC News 2021a). Even the president of Brazil's National Economic and Social Development Bank has acknowledged the problem:

BOX 3.6 THE INVERSE CARE LAW APPLIES IN THE GLOBAL SOUTH TOO

Health geographers are very familiar with the inverse care "law", introduced by general practitioner Julian Tudor Hart (1971), to assert that healthcare in Britain is provided least in poorer areas of greatest need. But the observation applies equally well in large parts of the Global South. Moosa *et al.* (2014) undertook interviews of health professionals trained in sub-Saharan Africa but who had not taken up posts in primary care in their own countries. Their findings make grim reading. For instance, a doctor from Rwanda says: "You know how to help them … but you can't really help them, because you do not have the resources." Another, from Gabon, reports: "It's, really, there is nothing. How can you treat people without a laboratory or without drugs, medication? No antibiotics, nothing." Others report lack of personal security. A nurse from South Africa observes: "I personally wouldn't want to work in … primary care in South Africa. I think you're exposed to so much, you know, so much danger." Another, from Zambia, tells us: "The most difficult … was when there was lawlessness." Geography matters too, as another South African nurse notes: "[Y]ou are having to go into certain of those areas, travelling miles to get to places, and then you've got all these cultural issues … and then you've got a whole language problem." Depressingly, the authors conclude that "effective primary care is not going to happen" (Moosa *et al.* 2014: e325; all the above quotes are also taken from this source).

For more quantitative evidence, consider Figure 3.7, showing the relation between child mortality (a measure of "need") and physicians per head of population (a measure of healthcare availability). The inverse relationship (using logarithmic scales) is very clear. The size of each country is proportional to its number of births.

Although the inverse care law plays out across geographical space, it happens in social space too. For example, Aboriginal women giving birth in South Australia report clear evidence of racial discrimination and unfair care (Brown *et al.* 2019).

GLOBAL HEALTH

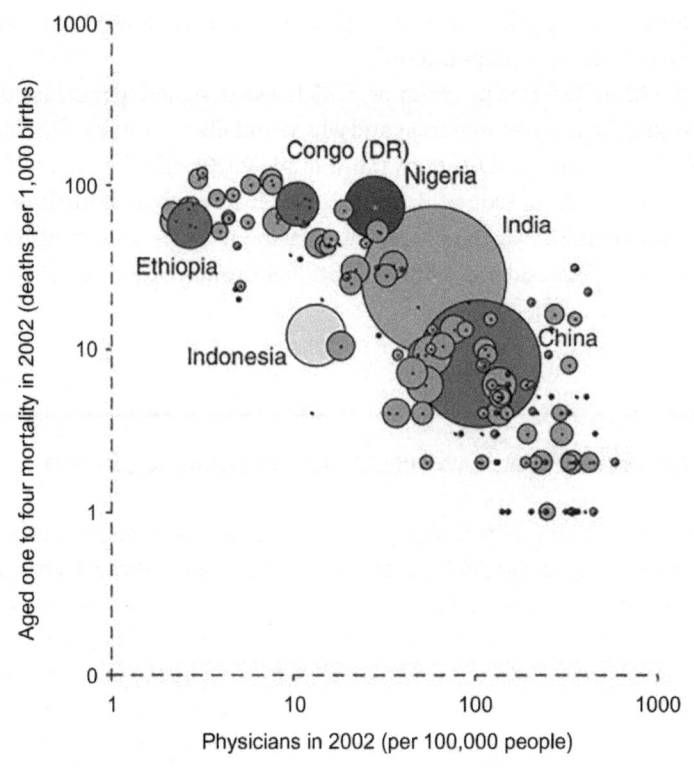

Figure 3.7 Association between child mortality and provision of physicians (2002)
Source: Dorling (2007) (courtesy of Danny Dorling, University of Oxford).

> For the elites, for the more developed classes, the crisis is gone. For the disenfranchised, the crisis will be rather long. It is further increasing our social imbalance, so we need to act with a sense of urgency, implementing tax and administrative reforms and continue the agenda of structural overhauls in Brazil.
> (AgênciaBrasil 2021)

As we shall see in Chapter 8, deaths from Covid-19 are socially and racially patterned in Brazil, as in other countries.

In India, Chatterjee (2017) reports that spending on public health is equivalent to only 1 per cent of GDP, leaving many people reliant on private providers. Access to and use of health services is spatially variable; states such as Kerala are relatively well catered for, but others, such as Uttar Pradesh, are not. In the village of Kheda, in Meerut district of the state, according to shopkeeper Arvind Som, one of Chatterjee's respondents, the primary health centre "is supposed to be open from 10am to 4pm but we have rarely found it to be so. I have seen a doctor for a few hours some days, but that is it". She continues: "[A] college student who needed to consult a doctor for an injury arrived. So did a man carrying a sick child. Both had to go elsewhere" (Chatterjee 2017: 2427).

The government introduced Mission Indradhanush in 2014 in order to increase levels of childhood immunization, particularly among hard-to-reach groups in rural areas. This has had some success. Bettampadi *et al.* (2021) use data from India's National Family Health Survey (NFHS) to show that the proportion of fully vaccinated children aged one to two years increased from 35 per cent in 1992/93 to 62 per cent in 2015/16, but they caution that the rate of increase has varied from state to state, implying that considerable inequalities persist. Similarly, inequalities according to socio-economic status remained; the authors report a significant gradient from the poorest to the richest quintile, in all states.

Sochas (2020) records that inequalities in access to adequate maternity services in low- and middle-income countries are particularly prominent, compared with other areas of primary care. With that context in mind, she explored, using GIS technologies, the links between geographical accessibility and social context in her study of maternity provision in Zambia, where the differences in high-quality provision between the wealthiest and poorest women are considerable. Barriers to care are not only geographical (distance from facilities); nonetheless, she finds that proximity to health facilities, along with the staffing and other resources required for high-quality care, is a significant determinant of safe delivery. A further study using GIS technologies reveals that Niassa and Cabo Delgado in the north of Mozambique are provinces that have the poorest access to primary healthcare facilities (dos Anjos Luis & Cabral 2016), reflecting other inequalities described in Chapter 2.

Mongolia is the 19th largest country in the world by area and the third most sparsely populated; almost half the population lives in the capital, Ulaanbaatar. Three-quarters of all maternal deaths are among herdswomen in remote rural areas (up to 213 per 100,000 live births, compared with 42 in the capital). Erdenee *et al.* (2020) examined the geography of midwife availability, and they note that the average distance from a rural household to the nearest primary healthcare facility is 100 km, and that accessing care requires overcoming barriers of cost and poor road infrastructure. In such a sparsely populated country, recording the number of midwives per head of population makes less sense than recording it by area. Results show that the fewer the midwives per area, the higher the maternal mortality rate; among the 18 provinces that had fewer than one midwife per 1,000 km^2, the maternal mortality rate was higher than the SDG target of 70 deaths per 100,000 live births. The unequal geographical provision has serious consequences for population health.

One possible solution to difficulties in accessing healthcare in LMICs lies in the use of electronic health (e-health) and mobile health (m-health) technologies. In many such countries there is extensive access to mobile phones, and in principle these could be used to access healthcare. However, there still requires to be a trained healthcare workforce and the financial means and political will to roll out these technologies. Kumar, Paton and Korigia (2016) suggest that such technology will improve access to healthcare services in these countries. They report that mobile phone penetration in Kenya was 88 per cent, but that uses for healthcare were hampered by disparities in the healthcare workforce, which ranged from 1.56 per 1,000 people in Isiolo county to 0.12 per 1,000 in Mandera county. One solution would be to draw on the diaspora of doctors and nurses located in high-income countries, particularly Britain and the United States, but it is unclear if this has taken off. Writing three years after Kumar, Paton and Korigia, Bervell and al-Samarraie (2019) find that, in Kenya, Botswana and Malawi, only three studies had explored the use of m-health or e-health for healthcare interventions. None were reported in the poorest countries, such as the Democratic Republic of the Congo, Sudan, South Sudan and the Central African Republic. Funding needs to be directed to human resources rather than information technology before m-health can begin to improve access to healthcare.

There is a wealth of studies showing that geographic accessibility to health services plays a significant role in shaping health outcomes. For example, children in northwest Ethiopia who lived more than 90 minutes from the nearest health centre were up to three times more likely to die than those living much closer (Okwaraji *et al.* 2012). In Burkina Faso a study of almost 3,500 child deaths shows that, allowing for confounding factors, walking distance to the closest health facility was significantly associated with infant and child mortality (Schoeps *et al.* 2011). Similar results are found for antenatal care. In Ethiopia, geographic distance is found to act as a constraint on visits to antenatal facilities (Tegegne *et al.* 2019). Of course, accessibility is not to be thought of only in terms of geographic distance. If a facility is closed when needed, or the cost of treatment is too high, it matters little if a child or expectant mother is on the doorstep.

Concluding remarks

This chapter has addressed seven key themes concerned with global health inequalities. Examining non-communicable diseases and their determinants in the Global South deflects attention away from the frequent focus on such countries as hearths of infectious disease (such as Ebola). Such determinants are not only social – poverty and inequality, most obviously – but also commercial, as large and influential corporations seek to extend their reach to new markets.

The discussion of gender inequalities reveals that women and girls in some countries in the Global South are not just subject to physical violence but are faced with the structural violence that comes from being poor. Gender and poverty intersect, most obviously for women struggling to balance low incomes with child-caring responsibilities, and seeing their own health often severely compromised. The health of Indigenous peoples contrasts with that of many of the majority populations, whether in North America, Australia or

New Zealand, and there needs to be a fuller appreciation of these inequalities, as well as an acknowledgement that the injustices and maltreatment suffered many years ago are still felt by today's native populations.

As with cities in the Global North, there are clear social and spatial inequalities in many urban areas in the Global South, with those living in overcrowded and poorly resourced settlements facing risks to health and wellbeing when clean water and adequate sanitation are not provided.

The burden of poor mental health across the globe is felt acutely in parts of the Global South and, as with NCDs, is shaped by poverty, income inequality and poor environmental quality. Those living in zones of conflict suffer the most, of course. Research on wellbeing in the Global North has mushroomed in recent years and is taking off in the Global South. First outlined in the United Kingdom, the inverse care law has been shown to apply in the Global South too. Access to health services is shaped by geographical distance as well as other factors, and poor and unequal access has serious consequences for women seeking maternal care.

All these themes illustrate the gross inequalities existing in the Global South – inequalities that are shaped by power relations and control of resources, whether by governments or particular social groups. Such issues of power lead us to consider in the next chapter how health is governed.

Further reading

On NCDs, see the World Health Organization's *Progress Monitor* (WHO 2020), but Herrick and Reubi (2017a) should be also consulted, as should Herrick, Okpako and Millington (2021) and Herrick and Reubi (2021). See also Amzat and Razum (2022: ch. 5).

On gender in health, you should be aware of the Women in Global Health collaborative, which seeks to redress the imbalances in leadership among healthcare professionals (see www.womeningh.org). A new edited collection (Williams & Luginaah 2022) is an important resource. Indigenous health geography is covered in the excellent overview by Richmond and Big-Canoe (2018); see also the references therein. I recommend too de Leeuw *et al.* (2012) and the overviews by Gracey and King (2009) and King, Smith and Gracey (2009).

A classic reference on urban health in the Global South is Harpham (2009). See also the collection of chapters in Mitlin and Satterthwaite (2012).

A key source on global mental health is White *et al.* (2017); see especially this chapter: McGeachan and Philo (2017). I also strongly recommend the comprehensive overview provided by Patel *et al.* (2018a). Patel and his colleagues (2018b) have also conducted a systematic review of the literature on income inequality and depression.

Allin and Hand (2014) have written on the measurement of both individual and national wellbeing. The World Happiness Institute produces annual global rankings of cities in terms of subjective wellbeing (see Helliwell *et al.* 2020). For an overview of wellbeing in health geography, see Severson and Collins (2018) and, especially, Atkinson (2016), and consult recent issues of the journal *Wellbeing, Space and Society.*

Last, on the provision of – and access to – healthcare, good general references in health geography are Barnett and Copeland (2010) and Ricketts (2010). There are numerous research papers on geographic access to healthcare in countries of the Global South. For a recent illustration using GIS technologies, see Wigley *et al.* (2020).

4
GOVERNING GLOBAL HEALTH

> The nature of global health is intrinsically political, reflecting power relations which play a significant role in whether various groups can set the political agenda and/or claim resources.   McInnes, Lee and Youde (2020)

This chapter looks at the remit of, and the activities engaged in by, different organizations and groups involved in global health. These include major world organizations (notably the World Health Organization and World Bank), constellations of public–private partnerships, non-governmental organizations and social movements. It also examines the role of philanthropy, asking whether it is more appropriate to speak of "philanthrocapitalism".

Wrapped up in the geopolitics of health are issues of global health security: issues that have inevitably come to the fore during the Covid-19 pandemic but that emerged some years earlier when new strains of influenza were appearing and the earlier coronavirus infection (severe acute respiratory syndrome: SARS) appeared on the world scene. The second half of the chapter therefore explores what is meant by health security.

As the quote at the start of the chapter suggests, we need to examine the relative power of the different agencies involved in global health, and where power lies in determining who gets "secured", how and where.

Governance

My focus here is on *relations* between the actors involved in global health – nation states and organizations of different types – and less on the governance *within* states or organizations. The latter matters greatly, of course. For example, there is a close relationship between maternal mortality rates among a set of 174 countries and indicators of their governance (government effectiveness, regulatory quality, rule of law, control of corruption, voice and accountability, and political stability and absence of violence – all indicators proposed by the World Bank) in those countries (Ruiz-Cantero *et al.* 2019). The stronger the governance, the lower the rate of maternal mortality.

As well as exploring the roles of nation states and major intergovernmental organizations, the governance of global health is seen as "stemming from a configuration of public private partnerships (PPPs) between governmental agencies and programs, NGOs and charitable foundations, including the omnipresent Bill & Melinda Gates Foundation and other "philanthrocapitalists" entwined with the neo-liberal order" (Lovell, Read & Lang 2019: 522). Each of these actors – and the relations between them – is discussed in this chapter. Relations can simply take the form of *communication* via the exchange of information (such as about the outbreak of a new infection). Actors might also *cooperate* or collaborate in dealing with an issue that cannot be tackled by a single one. Last, they can *coordinate*, leaving a single actor (an organization) to direct actions towards a common goal. As we see shortly, the World Health Organization was established as just such a coordinating body.

A key question is where the real power lies in shaping global health and global health security. Does such power lie in the hands of the nation state, or coordinating bodies such as the WHO? This is a question that concerns theorists of international relations. Globalization might suggest that nation states are ceding power to continental or global political institutions (the European Union or the World Bank, for example) that can exercise more authority than sovereign states acting in their own interest (Boyle 2021: 132). This has been resisted in some countries, as populist governments or political parties seek to (re) assert the autonomy of the nation state. The most telling recent example was the election of Donald Trump to the presidency of the United States in 2016, one of his early decisions (subsequently rescinded by President Joe Biden) being to withdraw from the WHO.

What, precisely, do major organizations such as the WHO do, and what relations do they have with nation states?

A world of organizations

Brown (2018) provides a valuable overview of the history leading to the establishment of the World Health Organization in 1948. It had been foreshadowed in the late nineteenth and early twentieth centuries by several intergovernmental agencies and international conferences, which established health regulations designed to respond to disease outbreaks. Many of these sought to address problems (of infectious disease) that lay outside a national border but within colonial territories. However, after the Second World War the newly established United Nations recommended creating an international health organization that would seek to improve health standards in all countries and enable better disease surveillance. "International" health became "global" health.

The early concerns of the WHO were with major infectious diseases such as polio, smallpox and malaria. Brown (2018: 258–9) discusses the latter in some detail, and he also notes the great success of wiping out smallpox. But, as many commentators have suggested, the focus on single-disease issues deflected attention from wider concerns with health prevention and primary healthcare, and these became the focus of the Alma Ata conference in 1978 (Box 4.1).

BOX 4.1 THE ALMA ATA CONFERENCE (1978)

In 1978 the 134 members of the World Health Organization met in Alma Ata (now Almaty) in the former Soviet Union (now Kazakhstan). The outcome of the conference – led by the Danish director-general of WHO, Halfdan Mahler – was to assert the importance of primary care and to call for a focus on the wider determinants of health: education, food and water, sanitation, immunization. One key declaration referred to the unacceptable existing *gross inequality* both between and within countries, suggesting that "upstream factors" were key determinants. Another of its ten declarations was that "*all countries should cooperate in a spirit of partnership and service to ensure primary health care for all people since the attainment of health by people in any one country directly concerns and benefits every other country*" (Pan American Health Organization [PAHO] 1978, emphasis added).

One further declaration at Alma Ata stated that "an acceptable level of health for all the people of the world by the year 2000 can be attained through a fuller and better use of the world's resources, a considerable part of which is now spent on armaments and military conflicts". As any reader knows, over 40 years later nothing has changed.

For a nuanced overview of the Alma Ata conference, see Rifkin (2018). For the WHO's statement of the Alma Ata declaration, see WHO (n.d.b). See also Brown, Cueto and Fee (2006) and Brown (2018: 259–62).

Not long after the conference it appeared that any aspiration for a comprehensive, "horizontal" approach to public health was unlikely. The WHO, headquartered in Geneva (Figure 4.1), returned to a "vertical" concern with specific programmes (of growth monitoring, oral rehydration therapy [ORT], breastfeeding and immunization – so-called "GOBI"). Basilico *et al.* (2013a: 102) suggest that ORT brings short-term benefits but is "only a Band-Aid solution to the lack of clean water and modern sanitation in many poor settings".

Subsequently, the debt crisis in the early 1980s spawned an increased role for the neoliberal approach of the World Bank and its manifestation in the form of structural adjustment programmes (SAPs), leading to what Basilico *et al.* (2013a: 91) call the "death knell of the primary care movement". Attempts by UNICEF to expand GOBI by adding family planning, female literacy and food supplementation (hence "GOBI-FFF") were only partially successful.

Figure 4.1 World Health Organization HQ, Geneva, Switzerland
Source: iStock.

Brown, Cueto and Fee (2006: 84) refer to 1988 to 1998 as a period of crisis for the WHO. The WHO has always received funding from two primary sources. One is the contributions from member states, assessed on the basis of their gross domestic product. The second comprises voluntary contributions, both from member states and the range of partners covered elsewhere in this chapter, such as philanthropic donations, the private sector and other UN organizations. In the late 1980s the proportions shifted dramatically away from member states towards extra-budgetary finance that was controlled by donations from wealthier countries and the World Bank. The latter's *World Development Report* in 1993 (World Bank 1993) referred to the WHO as a full partner but it was clear that the Bank controlled more of the purse strings. However, with a new director-general, Gro Harlem Brundtland, in place the 2001 WHO report "reaffirmed the role of the state in providing services in poor and middle-income countries, and cautioned against reckless privatization" (Brown, Cueto & Fee 2006: 89).

In 2018/19 the largest contributors to the WHO were the United States ($853 million), the United Kingdom ($464 million) and the Bill & Melinda Gates Foundation ($465 million). Interestingly, the budget for 2020/21 marked a "change from a disease-specific approach to a more integrated and health-systems-oriented approach to drive sustainable outcomes" (WHO 2019: 3).

Lovell, Read and Lang (2019: 522) claim that the politics of health shifted radically from the late 1970s through the present with an "economic turn within international health, and the health turn within development", as reflected by the World Bank. It is time to explore some of the other contributors to the WHO, starting with the role played by the World Bank.

The World Bank was established in 1944 – alongside a partner organization, the International Monetary Fund – to help rebuild the economies of postwar Europe. Briefly, the IMF provides temporary assistance to provide financial stability, while the World Bank became the key international body for economic development. In taking over as president of the World Bank in 1968, Robert McNamara shifted its focus to poverty reduction, with an initial stress on family planning, then nutrition, then maternal and child health. As Brown *et al.* (2018: 262) show, World Bank policy in the mid-1970s was to promote health using the public sector, referring – somewhat surprisingly – to distortions created by free markets. However, despite some successes in reducing levels of maternal mortality, subsequent leadership at the Bank asserted a neoliberal approach that held sway in the 1980s and 1990s. In particular, structural adjustment programmes were introduced (Box 4.2).

BOX 4.2 THE WASHINGTON CONSENSUS

Friedrich Hayek and Milton Friedman are the two Chicago economists most closely identified with neoliberalism, the ideology that became dominant in promoting free markets, and one adopted by the World Bank and IMF that was seen as "the only economic model capable of lifting Global South countries out of poverty" (Boyle 2021: 106). The World Bank and IMF joined with the World Trade Organization (WTO; see below) to impose conditions on "structural adjustment" loans they made. The principles on which these structural adjustment programmes were based became known as the "Washington Consensus" and included liberalizing (that is, opening up) trade, deregulation, privatization, prudent financial discipline and flexible labour markets – all key features of the late twentieth-century global economy. Put simply, SAPs required controls on government spending.

Many commentators, including those from within the organization, have criticized the World Bank's role in the 1980s and 1990s. For example, Joseph Stiglitz, former chief economist at the Bank, resigned in 1999, arguing that the economic reforms (the SAPs) it imposed on countries as conditions for loans were counterproductive and worsened the health prospects of their populations.

For further useful summaries, see Boyle (2021: 106–12, 215–20) and Brown (2018: 262–4).

Research has explored in detail the impact that SAPs have had on low- and middle-income countries. Thomson, Kentikelenis and Stubbs (2017) provide an overview of the impact of SAPs on child and maternal health, finding that the impact has been detrimental, weakening access to good and affordable care. A specific example is provided by Pongou, Salomon and Ezzati (2006), who document the impact of macroeconomic policy on child malnutrition in Cameroon. The government imposed cuts in public spending in the late 1980s, and drops in income per capita led to reductions in household spending on food and healthcare. Between 1991 and 1998 malnutrition in children aged under three years rose from 16 to 23 per cent on average, and the increases were higher in rural areas and in particular regions; in northern Cameroon malnutrition rose from 29 per cent to 34 per cent. Pandolfelli, Shandra and Tyagi (2014) find that those sub-Saharan African countries in receipt of a structural adjustment loan had higher maternal mortality rates than sub-Saharan countries that did not receive such loans. They suggest that, other things being equal, between 1990 and 2005 about 360 maternal deaths per 100,000 live births could be attributed to structural adjustment. More recently, Forster *et al.* (2020) have looked at the impact of SAPs on both neonatal mortality rates and health system access (a composite variable using data on antenatal care, immunization and other indicators) in 137 countries. Their results suggest that SAPs have worsened both access and neonatal mortality.

The role of the World Bank became, and remains, in part a health one. "In line with its global strategy for health, nutrition and population, the World Bank Group supports countries' efforts to achieve universal health coverage through stronger primary health systems and provide quality, affordable health services to everyone – regardless of their ability to pay" (World Bank n.d.a). But the extent to which it has helped provide "affordable health services to everyone" is highly questionable, since user fees are common and serve to cement inequity. The stated mission is to reduce the proportion of the global population living in extreme poverty to 3 per cent and "to promote shared prosperity by increasing the incomes of the poorest 40 per cent of people in every country" (World Bank n.d.a). Quite clearly, the World Bank has been deeply enmeshed in global health governance.

Schrecker (2020: 480) offers an excoriating perspective on institutions such as the World Bank. "Global health politics must be understood with the recognition that suffering can be inflicted and lives destroyed by remote control: by choices made half a world away, by people and within institutions that have no contact with those affected and are in no way accountable to them." He draws an uncomfortable analogy with drone strikes that inflict what is invariably labelled "collateral damage" when they miss intended targets.

The World Trade Organization, with 164 member countries, sets out a global system of trade rules and helps to negotiate trade agreements and settle trade disputes. Its mission states that "it supports the needs of developing countries" (WTO n.d.a), so we need to consider if it does so and how its work relates to global health. The WTO was established only in 1995 but was preceded by the General Agreement on Tariffs and Trade (GATT), which had operated since 1948. As well as covering trade in goods, the WTO concerns itself with trade in services, and with intellectual property (IP). It is the last of these that has proved most contentious for global health, since large pharmaceutical companies have sought to protect their patents for medicines (such as those for HIV and tuberculosis), keeping prices too high for those in need. The WTO Agreement on Trade-Related aspects of Intellectual Property Rights (TRIPS) is considered more fully in Box 4.3.

BOX 4.3 GETTING TRIPed UP

The Agreement on Trade-Related aspects of Intellectual Property Rights was reached in the mid-1990s, partly as a result of lobbying by the US pharmaceutical industry to protect its interests; but it covers many areas of the economy, including services, designs, performing arts, and so on. In particular, the pharmaceutical industry has long argued that the pricing of drugs should be set at levels that acknowledge costs of research and development. It seeks to protect its intellectual property by preventing companies that might otherwise produce similar generic (rather than brand-name) drugs and selling these at lower prices. Few LMICs are in a position to manufacture drugs to deal with common diseases; nonetheless, "big pharma" has continued to argue its case while poorer countries have sought waivers on patents. The so-called Doha Agreement in 2001 acknowledged the need to "promote access to medicines for all", and the WTO confirmed that the TRIPS agreement "allows low cost generic medicines to be produced and exported under a compulsory licence exclusively for the purpose of serving the needs of countries that cannot manufacture those products themselves" (WTO 2021).

The issue of waivers has become a major issue as a result of the Covid-19 pandemic. Patent waivers for vaccines have been blocked by the United Kingdom, European Union, United States, Germany and Switzerland. An attempt, led by India and South Africa, has been made to ensure that no barriers are placed in terms of timely access to affordable vaccines. Meanwhile, critics of patent waivers continue to argue that these set a precedent that may deter firms from investing in R&D. As of early 2023 there was still no firm agreement on a waiver for Covid-19 vaccines. Even before the pandemic the TRIPS agreement had been criticized for being overly stringent about the protection of intellectual property with regard to medicines – particularly those to treat HIV/AIDS and drug-resistant tuberculosis.

The pharmaceutical industry is considered more fully above (p. 61).

The same issues or tensions concerning national sovereignty versus the powers of a supranational body arise with the WTO as they did with regard to the WHO's position. To what extent can nation states protect their citizens' public health by restricting trade in products (such as tobacco or alcohol), when trade liberalization is a *raison d'être* of the WTO? When the Australian government sought in 2011 to introduce plain packaging for cigarettes the tobacco companies accused it of breaching international obligations.

The WTO says that TRIPS gives assurances to its members that they have "latitude to achieve their domestic policy objectives" (WTO n.d.b).

Bloche and Jungman (2006: 250) suggest that "portrayal of the WTO and its associated agreements as implacable threats to the health of people constitutes pessimism bordering on panic". In contrast to this rather dismissive tone, other research – particularly that promoted by those in favour of "deglobalization" – has indeed linked trade liberalization to worsening health outcomes, notably with respect to dietary change and the increase in obesity. Thow and Hawkes (2009) suggest that bilateral trade agreements between the United States and Central American countries have helped facilitate the "nutrition transition", with imports of foods high in sugar, salt and fats leading to rising rates of obesity as well as cardiovascular disease and cancer. In this sense, trade liberalization becomes another of the major upstream determinants of health. The World Trade Organization has tended to concern itself more with food safety rather than nutrition.

Public–private partnerships

Sustainable Development Goal 17 seeks to "strengthen the means of implementation and revitalize the Global Partnership for Sustainable Development". Explicitly, it "further seeks to encourage and promote effective public, *public–private* and civil society partnerships" (emphasis added). In terms of global health, public–private partnerships (PPPs) are alliances between private sector companies and public health. Frequently the private sector involves pharmaceutical companies, but PPPs have also involved companies in the food and alcohol industries, whose motives can clearly bring them into conflict with public health promotion.

There is more evidence to justify the use of PPPs in tackling infectious disease than non-communicable diseases. Here, I consider two examples of PPPs: first, the Global Fund to Fight Aids, Tuberculosis and Malaria; and, second, the Global Alliance for Vaccines (GAVI).

The Global Fund to Fight Aids, Tuberculosis and Malaria (often abbreviated to "the Global Fund") was established in 2002 to provide finance to prevent and treat those three diseases, such as by providing anti-retroviral therapy (ART), drugs to treat TB and insecticide-treated bed nets to prevent malaria. It claims to have saved 44 million lives by the end of 2020, and delivered ART to 21.9 million people, treated 4.7 million for TB and distributed 188 million bed nets in 2020. Donations come mainly from nation states in the Global North, primarily from the United States, United Kingdom, Germany, France and Japan. Some of its financing is outcome-based or result-based, with money transferring when pre-agreed outcomes or results have been achieved. The history of the Global Fund has been chequered, in part because of allegations of corruption and fraud (such as shipments of malaria medication being stolen for resale, and misappropriated funds), which led to reductions in donor pledges.

McCoy and Kinyua (2012) argue that, initially, the Global Fund sought to disburse funding quickly and pragmatically, and note that there was only a weak positive correlation between funding per head of population and disease burden. Latterly, a more needs-based approach has been developed. Consequently, for example, the Global Fund has supported the Regional Malaria Elimination Initiative (RMEI) in Central America, drawing on funds from the Inter-American Development Bank, the Carlos Slim Foundation and – inevitably – the Bill & Melinda Gates Foundation.

The Global Fund is also linked to PEPFAR, the President's Emergency Program for AIDS Relief, created under President George W. Bush in 2003. This forges "strategic partnerships with the private sector that support and specifically complement prevention, care, and treatment, addressing key gaps in innovative ways" (PEPFAR n.d.). But its partnerships also include the WHO and on-the-ground community organizations. Pfeiffer (2013) has examined the work of PEPFAR in Mozambique, where HIV is prevalent, albeit varying place to place within the country; current prevalence is 13 per cent among adults aged 15 to 49 years. PEPFAR has invested some $4 billion since 2004, including the construction of new hospitals to treat AIDS patients. But, as noted throughout this book, the upstream causes are neglected. Pfeiffer (2013:168) speaks of a "shadow epidemic" of hunger and poverty.

GAVI started in 2000 with the aim of increasing access to vaccines and immunization among those living in the Global South. Its funders are outlined in Figure 4.2, with major donor countries being the United Kingdom, United States, Norway and Germany, but very substantial resources coming from the BMGF. Private sector finance comes from vaccine manufacturers such as Pfizer, GlaxoSmithKline (GSK) and Johnson & Johnson.

GAVI has been criticized (by Médecins Sans Frontières [MSF], among others) for putting too much emphasis on marketing new vaccines (clearly, of benefit to pharmaceutical companies) rather than ensuring that known, effective, basic vaccination is universally rolled out (Birn 2014). As with other programmes seeking to improve health outcomes in the Global South, there are concerns that GAVI has adopted a vertical, disease-focused orientation rather than a horizontal one that would promote comprehensive and accessible primary care. However, since 2021 GAVI has committed to strengthening health systems and appears to be addressing this criticism. Recent research on equity gaps in immunization coverage suggests that GAVI is now focusing more widely on immunization access and uptake. Dadari *et al.* (2021) highlight India and Afghanistan as examples of countries that have adopted an equity lens (with respect to gender, income and religion, for example), though, sadly, one must question the likelihood of this focus remaining in Afghanistan.

Birn (2014: 14) observes that, "when PPP benefits such as direct grant monies, tax subsidies, reduced market risk, reputation enhancement, expanded markets, and IP rights are taken into account, the net result is that most PPPs channel public money into the private sector, not the other way around". But Craddock (2017) offers a more balanced

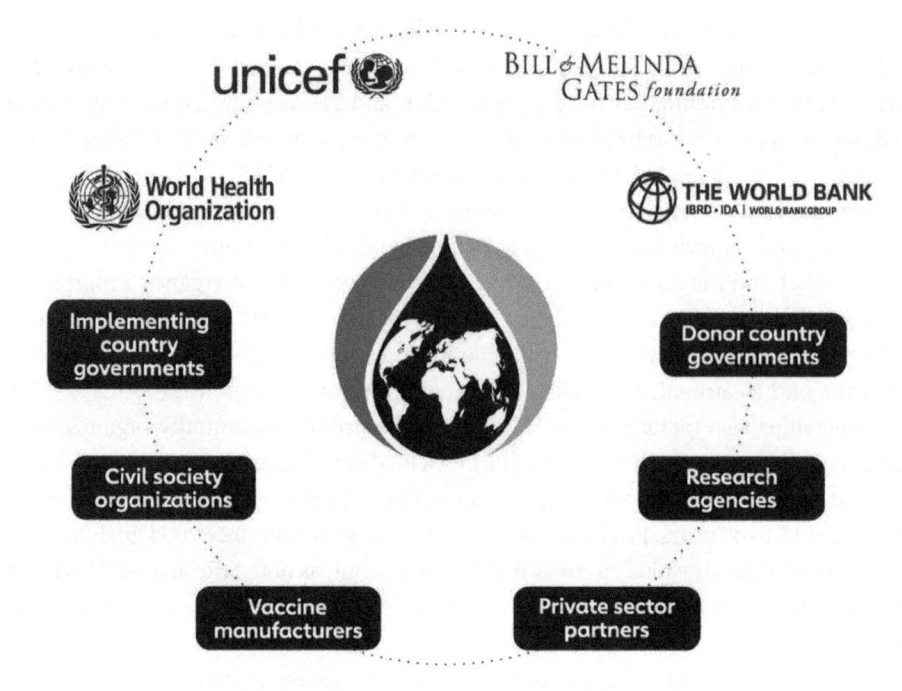

Figure 4.2 GAVI: Global Alliance for Vaccines
Source: Courtesy of gavi.org.

view of PPPs. Although she notes the critiques made that such partnerships prioritize technological interventions at the expense of broader public health measures, she does not see these as either/or. She suggests that they can bring wider benefits, such as providing infrastructure and training, for example on clinical trials.

Non-governmental organizations

A very broad definition of a non-governmental organization suggests it is any entity that is not state-based; however, this would include many of the multinational organizations such as the WHO and World Bank – or even public–private organizations – that have already been discussed. Here, I take a health NGO to be an international organization outside state control, funded externally. Examples include Partners in Health (PIH), Médecins Sans Frontières and Health Alliance International (HAI). Although these are separate organizations, as is often the case with global health partnerships there are close links between them.

The specific mission of Partners in Health – the brainchild of the late doctor-anthropologist Paul Farmer – is to focus attention on the poorest poor; a mission that is "both medical and moral, based on solidarity rather than charity alone" (PIH n.d.a). The "partners" in PIH include MSF, universities, the Global Fund, the WHO and national governments and the BMGF and other large foundations, as well as corporate partners such as GSK and Johnson & Johnson. "We believe lasting change is achieved through strong partnerships, which we have forged with medical and academic institutions, national governments, and our generous supporters" (PIH n.d.b). Its focus is on strengthening health systems, providing "well-trained staff; proper and ample medications and supplies; health facilities with reliable space, electricity, and running water" (PIH n.d.a). Spicer (2015), Pfeiffer *et al.* (2008) and others have criticized NGOs for their tendency to be disease-specific (focusing on vertical programmes), but the approach by Partners in Health suggests that it is adopting the horizontal approach that critics have cried out for.

Pfeiffer *et al.* (2008) offer what they call a code of conduct for NGOs engaging in health projects. They fear that the proliferation of NGOs has led to a loss ("brain drain") of key health workers from the public sector to these organizations. Pfeiffer and his colleagues report on their work with Health Alliance International, an NGO with its origins in Mozambique in 1987, that has done "what many global health NGOs have neglected to do: focus on strengthening public sector health systems by listening to and serving the needs articulated by countries and advocating for policies and practices that promote social, economic, and health equity" (HAI n.d.). In October 2021 HAI transferred all its responsibilities to be under the direct control of locally registered NGOs, in Mozambique, Timor-Leste and Côte d'Ivoire.

Others question whether the interventions of NGOs are emblematic of a former colonial, and paternalistic, ethos. Sakue-Collins (2021: 976) offers a trenchant critique, arguing that they are "(un)witting allies in the neoliberal network/project of arrested development in Africa". Gautier *et al.* (2022: 180) agree, arguing that NGOs risk perpetuating unequal relationships between donors and recipients; they illustrate this in their consideration of responses to Ebola outbreaks (Box 4.4).

BOX 4.4 THE 2014 EBOLA OUTBREAK AND THE ROLE OF MSF

Ebola infection is considered more fully in Chapter 8 but, in response to the 2014 outbreak, a panoply of international organizations, NGOs and other agencies came together to try to control disease spread in the three countries (Guinea, Liberia and Sierra Leone) most affected. There was limited public understanding of the disease and distrust of health authorities and workers, with widely held beliefs that Ebola was being spread by hostile governments. Setting up temporary treatment centres by NGOs helped perpetuate that perception (Gautier *et al.* 2022). The visibility of infection teams wearing "space suits", the perceived inhumane treatment of infectives and hurried burials did nothing to reassure local communities. As so often, wider health determinants were left unexplored. However, in acknowledging the need for interventions that are not marked by coloniality, Gautier and her colleagues acknowledge that help from Western technical experts is often required.

Unsurprisingly, there is a different perspective from MSF. A paper on its ethics framework (Sheather *et al.* 2016) recognizes that "parachuting innovations into complex environments without working collaboratively with affected individuals and populations can be perceived as patronizing, undermine trust, and result in failure [and] humanitarian interventions are often characterized by a significant power differential". The organization gives a specific – and geographical – example of practical assistance. They encountered difficulties in rapidly locating villages in Sierra Leone that were missing from maps but where cases of Ebola had been notified. Local people were given GPS-enabled mobile phones to gather information about the locations of villages, as well as the key local contacts. Data were entered into open-source GIS software to develop up-to-date maps of the district to enable monitoring of the outbreak.

Herrick and Brooks (2020) have considered the role played by international student medical volunteers in managing the outbreak in Sierra Leone, in part to bolster their emerging careers. As they note, not all humanitarian interventions are altruistic.

Philanthropy or "philanthrocapitalism"?

Philanthropy involves the voluntary transfer of money, whether from individuals or organizations, to improve the welfare of others. Although the Bill & Melinda Gates Foundation (Box 4.5) is currently the largest philanthropic donor organization in global health, there is a long history of similar bodies, such as the Rockefeller Foundation, which founded and supported the International Health Division in the early part of the twentieth century (see Birn 2014 for a lengthy description). And, despite the BMGF's pre-eminence, there are other significant contributors, such as the Wellcome Trust in the United Kingdom. In discussing the role of philanthropy, we should bear in mind ongoing contemporary debates about where historical figures drew their money from; many large projects and buildings were funded through profits made in the slave trade, for example.

Youde (2020) suggests that philanthropy introduced resources and innovations to global health. For example, although pharmaceutical companies may fail to invest in drug therapies when they see little commercial return, organizations such as the BMGF have stepped in to subsidize drug development. However, Youde also says that philanthropy raises questions about accountability. It also raises questions of power, whether over, or with, other entities.

BOX 4.5 THE BILL & MELINDA GATES FOUNDATION

Birn (2014: 16) suggests that the BMGF emerged "on the scene precisely at the apex of neoliberal globalization". It was established in 2000, and up to 2021 had disbursed £65.6 billion in grants (Gates Foundation n.d.a), making it a major player in global health politics. The BMGF portrays itself as working in partnership with governments, providing resources that governments cannot. As well as collaborating with governments, it supports a variety of public–private partnerships. For example, it has ties with the World Bank, the WHO, research organizations such as universities, and NGOs.

But its work has long adopted a technological/biomedical approach, funding potential treatments for particular diseases, notably HIV/AIDS, tuberculosis and malaria, rather than addressing more neglected diseases or providing the infrastructure that might be more sustainable for low-income countries. Birn (2014) suggests that larger-scale investment in horizontal primary healthcare would have paid greater dividends.

Mahajan (2018) has painted a different picture of the BMGF's influence. She suggests that its portrayal as having a huge influence in global health needs to be juxtaposed with the steady influence of a sovereign state such as India. Despite its early large investments in funding HIV prevention in 2002/03 it later engaged closely with the government, and funding for HIV transitioned to the state. As Mahajan (2018: 1366) puts it: "Studying the work of the Gates Foundation in India over a decade and a half, with a focus on its relations vis-à-vis the government, produces not a picture of hegemony but a more qualified sense of impact." We need, she argues, a more nuanced perspective on the relative balance of major philanthropic organizations such as the BMGF and the nation state.

The BMGF is described by its critics (such as McGoey 2016) as undemocratic, unaccountable and unanswerable to the public, a criticism that was made many years earlier about the Rockefeller Foundation. Birn (2014: 1) suggests, controversially, that both the Rockefeller and Gates Foundations were "started by the richest, most ruthless and innovative capitalist of [the] day".

This has led some writers to speak of "philanthrocapitalism" or actors as "philanthropeneurs", suggesting that the aspirations of donors do not match those of the low-income countries they seek to help. As with historical donors, one can argue that philanthropists are able to donate because they have businesses that have extracted wealth from exploited "others". The Gates Foundation has historically invested in fossil fuel companies that contribute to climate change, which itself impacts on poor health (see Chapter 9), including some conditions that the BMGF seeks to prevent.

In essence, philanthropy may attend more to downstream than upstream determinants of health because of a technological bias. Yet, although it is true that the focus is a technological one, the Gates Foundation does devote resources to improving sanitation and hygiene, albeit with a commercial focus. "Getting new sanitation products to market and into communities where they can save lives and protect community health will require the leadership of private companies that are excited about building new businesses in the sanitation sector" (Gates Foundation n.d.b).

In 2006 a new form of philanthropy emerged, in the form of "Product Red", the brainchild of U2 singer Bono. Established to contribute to the Global Fund (see above), it invites partner companies (such as Apple, Nike and Starbucks) to donate to the Fund proportions of their profits from specific products. Although Red claims to have distributed $150 million to the Global Fund, its critics argue that it spends vast amounts on advertising, lacks transparency and is "marketized philanthropy" (Birn 2014: 30).

Social movements

In a health context, civil society activism pushes for improvements in public health and addressing health inequities. A good example of such voluntary action is the People's Health Movement (PHM: www.phmovement.org), formed in 2000. This was a response to the failure of the Alma Ata Declaration in 1978 to secure "Health for All" by 2000 (Box 4.1). PHM has sought to bring together grassroots health activists across the world but with an emphasis on those from low- and middle-income countries. Its People's Charter for Health stresses that "[i]nequality, poverty, exploitation, violence and injustice are at the root of ill-health and the deaths of poor and marginalized people" (Narayan 2006). It has had some influence on WHO policy, notably in shaping its work on the social determinants of health. Musolino et al. (2020) see it as part of a counter-hegemonic social movement. They look at the work undertaken by 15 activists involved in the PHM. Some of these had become active during conflict or living under oppressive regimes; others had experiences of working with refugees, or came from families who were politically engaged. Some of those interviewed bemoaned the shift in power towards well-funded NGOs that could become disconnected from local communities. Activist Kate refers to "BONGOs, which are 'briefcase only NGOs', and basically where you can feel quite out of sorts because people are working for NGOs that have got no activist background whatsoever ... [T]hey are there because they can get an extremely good salary." The main tension reported by the activists was in seeking to balance local interests and global agendas.

More recently, the World Health Network has been formed as a coalition of citizens and experts seeking to address the global public health crises that have arisen because of Covid-19. The network is critical of many governments whose policy responses seem to accept ongoing community transmission and its consequences for long-term illness and mortality. A similar social movement, the World Social Forum, has a more general remit to contest growing neoliberalism and globalization. It contrasts with the annual World Economic Forum (WEF) held in Davos, Switzerland, giving a voice to those excluded from the global elite who attend the latter. Lee (2015: 41) refers to the WEF as yet another "key pillar of contemporary economic globalization".

In concluding this section we may contrast what I call the "global health governance optimists" with those taking a more sanguine view. Nunes (2014: 116) suggests that health can be a "bridge for peace", pointing to examples of regional collaborations, such as the Middle East Consortium for Infectious Disease (MECIDS), which links agencies in Israel, the Palestinian Authority and Jordan. The Mekong Basin Disease Surveillance Network, established in 1999, is another example. There is also an African Coalition for Epidemic Research, Response, and Training (ALERRT), which involves 13 African countries and "aims to reduce the public health and socio-economic impact of disease outbreaks in sub-Saharan Africa by building a sustainable clinical and laboratory research preparedness and response network" (www.alerrt.global). These networks may have wider political impacts that extend beyond health ones.

Kickbusch (2005) paints a somewhat optimistic future for global health governance, pointing to agreement on the SDGs and a shift in power dynamics away from US dominance and towards South–South cooperation. Gill and Benatar (2017: 240) are critical of Kickbusch's proposal that global health has the goal of "equitable access to health in all regions of the globe", suggesting that this "would only be remotely possible with an entirely differently structured global political economy involving understanding of and measures to address the interface between the ecological crisis, the financial crisis and the crisis of social dislocation".

Global health security: what is it and who is it for?

The World Health Organization defines global health security as "the activities required, both proactive and reactive, to minimize the danger and impact of acute public health events that endanger people's health across geographical regions and international boundaries" (WHO n.d.c). The WHO goes on to suggest that, with increasing population mobility and economic interdependence, global health threats of pandemics, environmental degradation, antimicrobial resistance (AMR) and climate change mean that traditional defences at national borders cannot offer sufficient protection.

Nunes (2014: 124) defines health security as "the existence of adequate mechanisms – in the form of legislation, resources and actors – to alleviate the structural inequalities that lead to harm and vulnerability to disease". This is all very well, but who authors the legislation, who provides the resources, who are the actors? And is health security only about disease?

If these definitions tell us what global health security is, we need to ask who it serves. Stoeva (2015) asks: is it security for the state, or for individuals? Any individual wants to have "secure" water, food, employment and to feel safe, as well as seeking good health.

Nunes (2014: 15) suggests that an emancipatory approach to security (the focus of his book) is concerned with the insecurities of "real people in real places" – engaging with those who have no voice. Further, alleviating insecurity and oppression can, he argues, address health inequalities. These arguments suggest it is the *individual* who is the focus. On the other hand, *national* security means protecting the state from external threats (whether aggression or the importing of new diseases), but it also covers internal threats (which might be challenges to government authority or public health challenges arising from obesity, for example). On the latter, as Brown (2011: 323) has observed, global health security invariably refers to infectious disease, not non-communicable diseases.

If health security is a concern of the nation state, surely there must be a role for the WHO in *global* health security? Benvenisti (2020) suggests that a founding principle of the WHO was to ensure that individual nations would *coordinate* their activities. It was, he argues, never the aim of the WHO to ensure international political *cooperation*, even though there has been scientific cooperation (most notably during the Covid-19 pandemic; see pp. 166–72 below). When the WHO was established it "was expected to primarily address health standards in developing countries; and, as a US-dominated body, it served (or was domestically presented as serving) Western efforts to demonstrate their commitment to the global poor against Soviet intentions" (Benvenisti 2020: 593).

Cooperation between states takes place more readily when there is equal vulnerability; this is demonstrably not the case in global health, nor has it ever been. It is therefore over-optimistic to assume that states will adopt whatever measures the WHO proposes to enable health security. The WHO cannot enforce; it cannot impose any sanctions on individual nation states. Indeed, revisions to the International Health Regulations (IHR) in 2005, designed to prevent, control and respond to the international spread of disease, reasserted state sovereignty. "The IHR both restricted the WHO's coordination function by limiting information and also undermined its cooperation-ensuring function by constraining the freedom of WHO leadership to act promptly vis-à-vis states' non-cooperative behaviour" (Benvenisti 2020: 596).

There is, therefore, a clear tension between sovereign nation states and a body such as the WHO. Writers on international relations (IR) theory clarify that this is the difference between "Westphalian" and "post-Westphalian" governance. This refers to the Treaty of Westphalia in 1648, which symbolized the principle of non-interference in the domestic affairs of nation states. By "post-Westphalian" is meant the primacy of the global over the national; the WHO is emblematic of such. However, despite the intensification of globalization and greater interaction (migration, trade) across borders, national interests and policies remain dominant (Stoeva 2015). This is certainly true of responses to disease threats. The pursuit of common goals for global health has been elusive, as I reveal in Chapter 8 in a discussion of responses to Covid-19.

BOX 4.6 GLOBAL HEALTH SURVEILLANCE

In 2002 Nicholas King noted: "Recognizing that physical sanitary cordons are impossible in a putatively borderless world, the emerging diseases worldview idealizes 'informational cordons', which would identify and manage risks before they become epidemics that threaten American citizens and interests" (King 2002: 773). Two of these "informational cordons" or major global surveillance networks have been established.

The Global Outbreak Alert and Response Network (GOARN: https://goarn.who.int/#aboutus) was formed in 2000, under the auspices of the WHO (Elbe 2010: 57–8). The intention of GOARN was to engage the resources of technical agencies beyond the United Nations for rapid identification, confirmation and response to public health emergencies of international importance. However, closer examination suggests that it is less of an alert system than a response one, responding to acute public health events simply by deploying staff to control disease spread. GOARN provided support during the SARS outbreak and has set up field laboratories for managing cholera, Ebola and other disease outbreaks.

The Global Public Health Intelligence Network (GPHIN), developed in Canada in 1997, is an informal, Internet-based early warning system that scans for news of outbreaks, bypassing sovereign nations (Weir & Mykhalovsky 2006). In 2009 agreement was reached with the WHO to use it to verify reports through its regional offices and country contacts; daily reports are produced and disseminated (Elbe 2010; Dion, AbdelMalik & Mawudeku 2015). GPHIN helped to flag SARS in 2003 and the H1N1 flu virus in 2009, as well as Zika and Ebola. However, according to reports in the Canadian press (Brewster 2021), its role in the detection of SARS-CoV-2 (the virus leading to Covid-19) in late 2019/early 2020 was apparently limited, since public health priorities in Canada had shifted elsewhere.

IR theorists have debated at length whether approaches to previous global health concerns (notably SARS) have been Westphalian or post-Westphalian. Fidler (2004) has argued that the WHO assumed independent authority during the outbreak, dictated policy to states and ushered in a new post-Westphalian era. However, Davies (2008) disagrees, suggesting that the WHO's role in SARS has been overstated. As she puts it, the "obstacle of sovereignty" has not been overcome (313), since the WHO relies on individual countries to cooperate.

Feldbaum *et al.* (2006), in their comparison of US and UK security strategies, note the tendency to construct as security threats those infectious diseases that might pose a possible risk to countries in the Global North, rather than those that have long posed a real threat to countries in the Global South. This point has been made by many others writing on global health security. Brown (2011), for example, notes that neglected tropical diseases, as well as non-communicable diseases, are generally absent from consideration as global health security threats; see also Sparke and Anguelov (2012). Despite Covid-19 being a true global pandemic, the *total* number of deaths (estimated in April 2023 as 6.91 million) compares with an *annual* total of over 3 million *children* dying from hunger and malnutrition.

Elbe (2010: 33–53) has traced what he calls the "medicalization of security". In chronological order, HIV/AIDS was the first among a set of emerging diseases to be linked to national security (King 2002), followed by SARS in 2002/3 and then avian influenza (H5N1). AIDS was seen as a security issue in part because of the risks of destabilizing societies and economies in the Global South. SARS and H5N1 (discussed further in Chapter 8) were constructed as economic threats, prompted by the growth of international air travel and the increased connectedness of places. Elbe (2010: 31) refers to a "global epidemiological space" created by air travel. Ten years before Covid-19, he refers to H5N1 as "the pandemic laying [*sic*] in wait", and he reviews briefly the response to both this and to SARS, via the creation of global surveillance mechanisms (Box 4.6) – a "new spatial organization of the geopolitical borders of infectious disease control" (Weir & Mykhalovsky 2006: 259).

The security of the nation state inevitably requires the securing of its borders, a theme to which I return in both Chapters 5 and 8 (on population movement and infectious disease). But, although attempting disease containment at the border matters, this does not deal with the upstream causes of infectious disease in the Global South and the everyday structural violence experienced by much of the world's population.

These strictures have not inhibited a group constructing a "Global Health Security Index" (Box 4.7). This index has been reviewed by Boyd, Wilson and Nelson (2020), who note that, although the index stresses "political, socioeconomic and environmental vulnerabilities as factors that can amplify deficiencies in preparedness", it does nothing in practice to account for these.

BOX 4.7 THE GLOBAL HEALTH SECURITY INDEX

The GHS Index seeks to measure the health security ("factors critical to fighting outbreaks") across 195 countries using data on 37 indicators and six categories: prevention, detection, response, health systems, commitments and risk environment. In 2021 the most "health-secure" countries were, in order, the United States, Australia, Finland, Canada and Thailand, whereas those rated least secure were Somalia, Yemen, North Korea, Syria and Equatorial Guinea.

As noted, the United States appears as the most secure country, but the 2021 report by the GHS group (Bell & Nuzzo 2021: 14) castigates it for having squandered its capacities for responding to the Covid-19 pandemic. More broadly, it invites countries to "prioritize the building and maintaining of health security capacities in national budgets" (Bell & Nuzzo 2021: 17), to identify risk factors and develop plans accordingly and to "conduct comprehensive after-action COVID-19 pandemic reports so that they can learn from this crisis and ensure that capacities developed during the pandemic are expanded and sustained for future public health emergencies" (17). These invitations seem, to put it mildly, hopelessly unrealistic for most countries in the Global South, and, despite the transparency with which data are collected and reported, the overall value of the GHS Index is questionable. Alarmingly, the 2021 report asserts that "countries are not prepared to prevent globally catastrophic biological events that could cause damage on a larger scale than COVID-19" (16).

See www.ghsindex.org and Boyd, Wilson and Nelson (2020) for more details, as well as the World Economic Forum's annual risk reports. Lakoff (2017), writing before Covid-19 emerged, discusses the (lack of) preparedness for global health emergencies.

Concluding remarks

What do we conclude from this discussion of governance and security?

First, despite the plethora of organizations that profess to manage global health, the role of the nation state has hardly diminished, as several writers confirm. Ricci (2009: 2) argues that, "despite claims of globalization altering the political landscape such that nonstate actors ... find themselves on more equal terms with the state, in fact the state remains the clear driver of international as opposed to global health policy". Schrecker (2020) agrees that, despite the presence of international organizations, decision-making in those organizations is controlled by the most powerful member countries. As an aside, he also observes that a nation state is not comprised of homogeneous interests; for example, in India the country's states wield considerable influence themselves. Finally, and with regard to Covid-19, Herrick and Reubi (2021: 3) note that the pandemic

> has closed the world in on itself as national containment strategies have shut borders and vaccine nationalism has pitted country against country. Amid this, national health has been cemented as the chief frame of reference even while the pandemic has underscored just how crucial it is to envisage (and manage) health at the scale of the global.

The WHO itself acknowledges this. An editorial co-authored by a director of a WHO Collaborating Centre notes that the WHO lacks the authority and the means to engineer an effective response to global health emergencies (Gostin, Moon & Meier 2020).

Second, in acknowledging the large and diverse set of organizations involved in global health, it is appropriate to question the leadership of these and their source of expertise. Admitting that it is a generalization, Reidpath and Allotey (2019) assert that the WHO derives its expertise mainly from relatively few places in the Global North. They argue that more training needs to be provided by institutions in the Global North to help those working in the Global South. Leadership tends also to be a male enterprise. Davies *et al.* (2019: 601) want to see feminist leadership but say this requires "more than gender quotas: it requires formal and informal cultural change within institutions across all areas of global health governance". Herrick (2017a) is concerned that the global partnerships undertaken by those working in the United Kingdom do not necessarily match where the needs are greatest, relating instead to the country's broader geopolitical interests.

Third, there is a danger that too great a focus on security deflects attention from the structural causes of poor health. Protection from infectious disease should be seen as an issue of human rights, not merely one of security. Relatedly, we hear little from those whose security is most threatened: not those in the Global North worried about the low risks of (for example) Ebola but those lacking secure food and water or personal safety in zones of conflict or want in the Global South. We need their stories to help understand the impact of global health governance.

Further reading

On global health governance and security, although much has changed since the emergence of Covid-19, there are many sources worth consulting. For a valuable geographical contribution, see Brown (2011), but the collection in Rushton and Youde (2015) is an essential resource, as are McInnes, Lee and Youde (2020) and Price-Smith (2009). Wenham (2021) has written on global health security from a feminist perspective, with a particular focus on Zika virus.

On structural adjustment programmes, it is worth reading Breman and Shelton (2006), who present a less critical overview than discussed above. Birn (2014) has a lengthy discussion of the history of philanthropy in global health, which sets this in the wider context of the multiplicity of actors involved; it is well worth reading.

Amzat and Razum (2022: ch. 8) discuss global health politics and diplomacy.

5
PEOPLE ON THE MOVE: THE DISPOSSESSED AND THEIR HEALTH AND WELLBEING

> Demagogues and national security experts now look askance at many of those who move, defining mobility as dangerous and threatening, while immobility is seen as normal and necessary for political and personal security.
>
> Glick Schiller & Salazar (2013: 184)

I noted in Chapter 1 that the essence of globalization lies in the various interconnections and interdependences across borders. These networked connections involve mobility, whether of goods, money, services, information or people. Such flows are the focus of this chapter and two of the next. In the present chapter the focus is on movements of people and the associated health impacts. Later chapters examine movements of materials – focusing less on the trade in goods and more on the flows of "bads" – and then infections.

In a very prescient observation ten years ago, Glick Schiller and Salazar (2013: 184) wrote that "the current global economic crisis seems to be accompanied by a normalisation – once again – of national borders and ethnic boundaries, even as the crisis itself reveals the degree to which the world is intricately networked and interdependent". The connections afforded by globalization seem – after the Russian invasion of Ukraine, as well as Trumpian politics and Brexit – to be giving way to heightened nationalism. Despite these borders, desperate people will continue to seek safety and refuge by crossing them, dangerous as the journeys usually are and uncertain as to the welcome – or lack thereof – that awaits them. The welcome given to Ukrainian refugees in Greece contrasts with the appalling conditions in camps "housing" Afghan refugees on the island of Lesbos (Markham 2022).

A key theme in contemporary social science is that of "mobilities". This is a vibrant multidisciplinary field that deals with the geographical mobility of humans, non-humans and objects, as well as information, visual images and, crucially, capital (Sheller 2015). It reflects more than the mere act of movement or transport, looking also at the emotional work that goes into this as well as the physical infrastructure necessary to enable movement. A related concept is that of "motility" – the potential to move – which "captures the unequal distribution of power over mobility, mobility resources and access to space" (Stjernborg, Wretstrand & Tesfahuney 2015: 384). The initial focus in the mobilities literature was, it is fair to say, more concerned with those moving by air, automobile and other forms of travel – the global elite – in the Global North. As Glick Schiller and Salazar (2013: 186) put it: "Those caught up in the initial exuberance of the new 'mobilities studies' not only understated the degree to which the poor and disempowered find themselves contained but they also projected movement itself as liberating, valuable and the basis of a new contemporary cosmopolitanism."

As revealed by some of the literature discussed below, the focus has shifted towards a more explicit concern with injustice and the mobilities of the dispossessed. In keeping with the overall theme of the book, my consideration of people on the move – clearly, a geographical matter – stresses the health and wellbeing of those who are not moving by choice. (For a wider consideration of the relationship between migration and health, see Gatrell & Elliott 2015: ch. 10).

Refugees, and those seeking asylum (Weber and Pickering 2011 prefer the label "illegal*ized* travellers"; note the italics), move not only from the Global South to North but within the Global South as well, while modern slavery, in the form of human trafficking for sexual exploitation and forced labour, is predominantly from south to north. Conflict in the Global North, such as the invasion of Ukraine by Russia in 2022, as well as upheaval in the wake of the Second World War (Gatrell 2019), has also led to huge numbers of refugees seeking safety further west. The impacts of such movements, including consequences for health, will be felt for years to come.

Other flows within and from the Global North, where people with more agency can afford to move with their work or spend vacations in warm climates, carry none of these appalling burdens. There, the movements of some enable the movements (and health outcomes) of others. "It is the labour of those whose movements are declared illicit and subversive that makes possible the easy mobility of those who seem to live in a borderless world of wealth and power" (Glick Schiller & Salazar 2013: 188). Migration for many can lead to entrapment rather than liberation.

Population movement can, of course, take place within nation states: dispossession and disadvantage occur there too. I look here at the health impacts both of flows across and within borders, with a clear focus on those who are displaced. Displacement can be forced for a number of reasons. I defer to Chapter 9 a discussion of the health impacts of population movement that is triggered by climate change. Here, my emphasis is on migration forced on people by political conflict and upheaval, with some attention given to movement driven by large-scale infrastructure projects and environmental hazards. My focus is very much on the Global South.

The scale of forced displacement

Those forcibly displaced from home fall into different categories. Briefly, a refugee is a person living outside the country of their nationality because of war, or a fear of persecution as a result of race, religion or political views. An "asylum seeker" is one seeking to be classified as a refugee but whose status has yet to be agreed. The United Nations High Commissioner for Refugees (UNHCR) estimates that there were 82.4 million forcibly displaced persons in 2020 (Table 5.1). Note that the number of internally displaced people is almost double the number of refugees and asylum seekers. The total number of refugees (26.4 million) represents about 10 per cent of the world's total of international migrants in 2020.

The countries with the most *internationally displaced* persons in 2020 were Syria (6.8 million), Venezuela (4.9 million), Afghanistan (2.8 million) and South Sudan (2.2 million). The countries hosting the most refugees (with the main countries of origin) were Turkey (Syria), Colombia (Venezuela) and Germany (Syria). The UNHCR points out that there were almost 1 million children born in displacement between 2018 and 2020, most at risk of permanent exile.

The countries with the most *internally displaced* persons (IDPs) in December 2020 are shown in Figure 5.1. The large number of IDPs in Colombia has tended to be rather unreported, certainly in comparison with those in Syria or the Horn of Africa. Many have been displaced from rural areas, often more than once, to informal urban settlements. This is driven by rural poverty but also military conflict, despite a peace agreement reached in 2016 between the Colombian government and the revolutionary group FARC.

There is a wealth of data on such displacements provided by the Internal Displacement Monitoring Centre (IDMC), which is based in Geneva. The IDMC has developed an Internal Displacement "Index" (IDI), which it uses to monitor the level of such displacement (Box 5.1).

Table 5.1 Global forced displacement, 2020

Category	Number (millions)
Refugees under UNHCR mandate	20.7
Palestinian refugees under UNRWA mandate[1]	5.7
Internally displaced people	48.0
Asylum seekers	4.1
Venezuelans displaced abroad[2]	3.9
Total forcibly displaced persons	82.4

Notes: [1] UNRWA = the United Nations Relief and Works Agency for Palestine Refugees in the Near East; [2] the UNHCR highlights this as one of the world's current large displacements, a result of violence and insecurity, as well as shortages of food and medicine.
Source: UNHCR (2022).

GLOBAL HEALTH

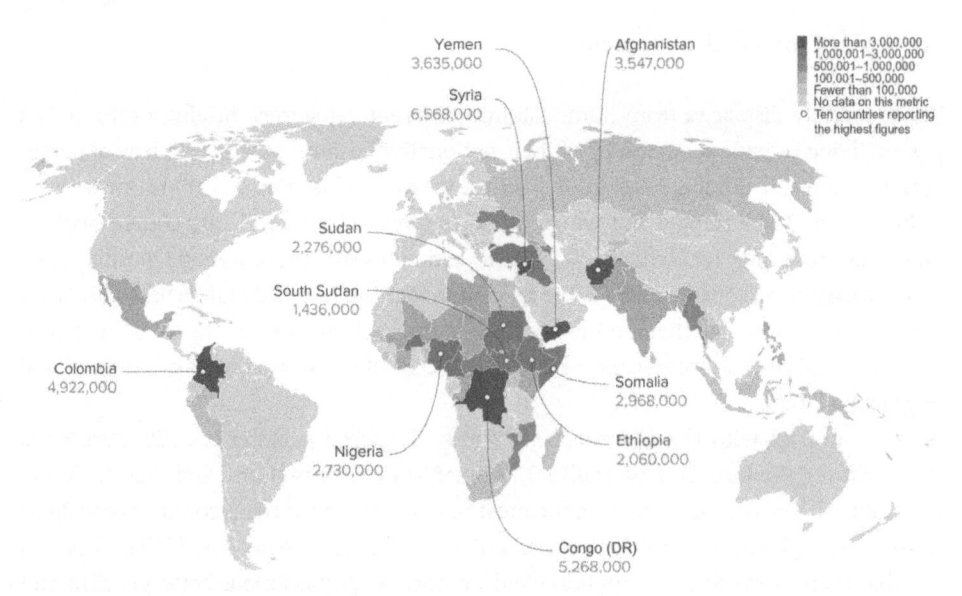

Figure 5.1 Number of internally displaced persons, December 2020
Source: IDMC (2021b).

BOX 5.1 THE INTERNAL DISPLACEMENT INDEX

The Internal Displacement Index is based on three components: context; policies and capacities; and impact. The context includes the political, socio-economic and environmental factors that affect displacement; policies and capacities refer to the resources available to prevent and manage displacement; and the impact includes the scale, severity and economic cost of internal displacement. Oddly, the index does not appear to measure the health impacts. Somewhat counter-intuitively, higher values of the index mean that countries encounter fewer factors to drive displacement, have reasonable capacity and experience relatively low impact. Yet the IDI figure for Colombia is 0.691, which seems curious given the data shown in Figure 5.1, and the values for Sudan and South Sudan are, respectively, 0.911 and 0.712. Syria (0.320) is among the lowest values.

The usefulness of such an index is questionable. It seems to offer a spurious degree of precision, and one must ask what value it is (to the international community) to know that a country with an index of, say, 0.451 is doing better than one measured as 0.441. More value is surely to be had by hearing the stories of those individuals who are displaced.

The 2021 report (IDMC 2021a) describes the results for 46 countries and provides maps of each component making up the index.

Bayar and Aral (2019) have sought to model the determinants of forced migration among a set of 48 countries in Africa. Unsurprisingly, their results suggest that poverty and conflict are triggers for forced migration, but whether we need a statistical model to confirm this is a moot point. A structuralist argument is more appealing, as proposed, for instance, by Weber and Pickering (2011), who suggest that much forced migration from sub-Saharan Africa and elsewhere is driven by the economic insecurity that is embodied in the relations between countries of the Global South and Global North. As an aside, the modelling by Bayar and Aral suggests that climate change has an indirect effect on forced migration; its role, and its health impacts, are considered more fully in Chapter 9.

The journey

The very word "displacement" has obvious geographical connotations: it means removal from the usual or "proper" place. Thinking of displacement as an assemblage (above, pp. 8–10), it is clear – particularly for those moving across borders – that it involves not only the individual or family on the move but a network of other human actors (relief personnel, border officials, criminal gangs) and non-human actors (documents, transport infrastructure, mobile phones) that combine together to enable – or thwart – mobility (Davies, Isakjee & Obradovic-Wochnik 2022). In addition, people will carry memories of home, and anxieties about future belonging, that will impact on mental health and wellbeing.

Forced migration, particularly that crossing international borders, involves periods of fragmented mobility and immobility. It is not a question of moving directly from home. Iranzo (2021: 740), in a study of migrants from sub-Saharan Africa to Spain, suggests that individuals and family groups may live itinerant lives for several years. Such itinerants may be extremely vulnerable, lacking in basic resources such as food and water, and threatened with violence (Gill, Caletrío & Mason 2011).

It is difficult for those living in the Global North to always appreciate the uncertainty and trauma involved in seeking refuge away from war and conflict. Many stories go unreported and untold (but see Box 5.2 for one example), although the hazards encountered by desperate people trying to cross the English Channel to reach the United Kingdom bring some of this trauma into sharp relief, regardless of whether one believes such movement should be discouraged by all means possible – the view of the current Conservative British government (Box 5.3). Of course, this is just one example, given oxygen by the popular press, which demonizes those contributing to the "flood" of migrants threatening the way of life of host populations. Other examples include those trying to cross from Mexico to the United States, risking drowning in the Rio Grande, or trying to cross the Mediterranean into Spain or Italy from north Africa. Australian writers Weber and Pickering (2011) consider "deaths at the global frontier" in considerable detail, although

BOX 5.2 FLEEING SUDAN

Mohammed (an alias), aged 25 years, left Sudan in January 2019 after Janjaweed militia massacred many of his fellow villagers. He spent a year in Libya under the control of traffickers before escaping by boat to Malta. From there he made his way through continental Europe, sleeping rough, before arriving in Calais in 2021. He crossed the English Channel to Dover in May 2022 after several abortive attempts, with ten other Sudanese refugees. He is now one of the first refugees facing forced removal to Rwanda, under the British government's proposed relocation scheme. He told a reporter (Taylor 2022): "I wanted to reach the UK because I heard it was a place where I could be safe [but] I haven't been able to sleep since I arrived here. My trauma is getting worse and I've been having more flashbacks about what happened to me in Sudan and Libya."

The British Home Office champions its "world-leading migration partnership" with Rwanda, arguing that it "means those making dangerous unnecessary and illegal journeys to the UK can rebuild their lives" in Rwanda (Home Office 2022).

---

their definition of "border deaths" is sufficiently broad to include those occurring en route as well as at the physical border. Crucially, they are more concerned with *accounting* for deaths, not merely *counting* them; although the latter is needed, it remains the case that many lives lost will remain unrecorded. Covid-19 had an impact on the modes of transport sought out by desperate people; the reduction in freight traffic (by lorry or ferry, for example across the English Channel) meant that far riskier crossings by unsafe boats were sought out, often at huge financial, not to say personal, cost.

BOX 5.3 ATTEMPTING TO CROSS THE ENGLISH CHANNEL

Refugees have sought to cross to England from France either by boat or on trucks/lorries using ferries or the Channel Tunnel.

The London-based Institute of Race Relations (IRR) has detailed about 300 deaths from attempts since 1999 to cross the English Channel. The institute's vice-chair asserts (Webber 2020) that such deaths are not simply tragic accidents but, rather,

> man-made, created by policies which do not merely close borders but also erect ever more obstacles to safe travel for the most vulnerable. Military-style solutions don't solve humanitarian problems. They simply create more profit for the smugglers, and more suffering for the migrants. The history of the securitization of the English Channel is a history of death.

PEOPLE ON THE MOVE: THE DISPOSSESSED

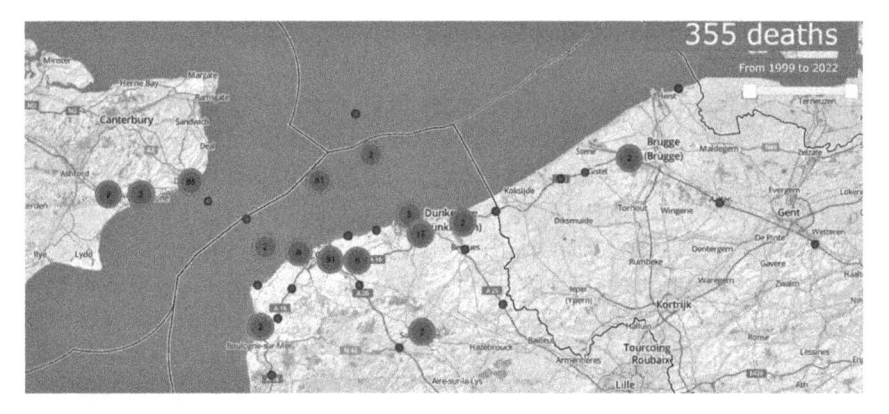

Figure 5.2 Mapping deaths of people trying to cross the English Channel, 2019–21
Source: https://neocarto.github.io/calais/en (courtesy of Maël Galisson and Nicolas Lambert).

The response of the United Kingdom's Home Office has been to introduce patrols by the Royal Air Force, and drone monitoring, but it has also suggested a physical blockade, laying nets to trap boats, or even the setting up of floating detention centres. Weber and Pickering (2011) draw on the concept of structural violence to illuminate the impact of border control policies such as these.

The French writer Maël Galisson has named (and in some cases photographed) many of the 300 deaths referred to above, including the 39 people found dead in the back of a refrigerated truck in an industrial park in Essex. An interactive map in his report (Galisson 2020) provides an opportunity to read, briefly, the stories that lie behind some of these deaths (Figure 5.2).

Any attempt to count the numbers of those who have died making this and other crossings runs the risk of obscuring the real and very personal human tragedies that underlie body counts. To give one example from August 2020:

> The body of 22-year-old Abdulfatah Hamdallah was found on Sangatte beach. He drowned after trying to cross the Strait of Pas-de-Calais on a makeshift boat. A native of west Kordofan, Sudan, Abdulfatah fled his country in 2014. He spent at least two years in Libya before reaching Europe. Adbulfatah asked for asylum in France in 2018 but his application was rejected. (IRR 2020: 40)

To offer a second example: in Calais "a makeshift graveyard has been set up to remember each of the refugees who have died trying to reach Europe. One of the wooden crosses has been placed for newborn Samir Khalida, who was killed when her mother, who left Eritrea for a better life in Europe, fell from a truck, triggering her premature birth" (Sharma 2021).

The health of those internally displaced

While the world's attention has, since early 2022, been understandably focused on Ukraine, this has deflected attention from other appalling situations. One example, that of the Rohingya minority in Myanmar, is discussed in detail in Box 5.4. Another is that of those remaining in Syria, 11 years after the civil war began. As reports by the Norwegian Refugee Council (NRC) suggest, the "daily routine remains the same: going to bed hungry, queuing up for hours to get hold of a bag of bread, and burning toxic plastic to stay warm" (Bayram 2022). The NRC tells in some detail the story of a family of Syrian Kurds displaced by conflict within Syria, travelling across the border to Lebanon, then going back to different parts of Syria, before ending up in Iraq, "helped" by smugglers across all borders. As the report says, it's not only the war: poverty, lack of access to medical care and lack of documentation all contribute to appalling mental ill health. The impacts of displacement are felt unequally, with women bearing a considerable burden because of childcare responsibilities. An overview of the health of internally displaced women in Africa by Amodu, Richter and Salami (2020) notes that women may be forced to have sex in order to provide food for themselves and their children, thereby exposing them to risks of infection. The authors refer to studies in Ethiopia and the Democratic Republic of the Congo, suggesting that such violence takes place in all kinds of public places – water collection points, sports fields, schools and markets – with perpetrators being not just the military and police but community members as well.

Evidence suggests that the health impacts of internal population displacement endure. For example, research on older Palestinians who had been displaced in 1948 (on the creation of the state of Israel) has revealed that their self-reported health was significantly worse – after controlling for socio-economic status and other factors – than that of people who had not been displaced; the scars are long-lasting (Daoud *et al.* 2012). The same is true of Indigenous groups removed from native lands onto reservations; as noted in Chapter 3, the trauma of dispossession is transmitted across generations.

Hay, Skinner and Norton (2019) review the impact of dam construction on internal population displacement. They suggest that between 2011 and 2020 the number of people displaced through infrastructure projects, including dam construction, was about 200 million over the decade. One major project, the Sardar Sarovar Dam in India, has been the focus of controversy for many years because of the inhumane way in which the displaced population has been treated (Roy 1999). Vulnerability is not evenly distributed: rural dwellers, farmers and indigenous groups are particularly at risk, and, when resettled, people (women in particular) are found to be at risk of violence (Hay, Skinner & Norton 2019).

BOX 5.4 ROHINGYA REFUGEES IN BANGLADESH

The Rohingya people – mainly Muslim – come from the state of Rakhine in Myanmar (a mostly Buddhist country) and have long suffered persecution, being deprived of citizenship and therefore rendered stateless. Almost a million fled to neighbouring Bangladesh (to a district called Cox's Bazar) in 2017 because of the systematic discrimination they had long faced, followed by government policy of ethnic cleansing (the US government has referred to this as genocide: Hern 2022). At present about 100,000 Rohingyas are internally displaced, in camps in Myanmar, and there are thought to be some 1.1 million Rohingya people in Bangladesh (Hossain *et al.* 2019).

The health status of the Rohingya refugees has been the subject of several research studies. Evidence suggests that children, who make up more than a half of the refugee population, have lacked basic immunization; half of them are missing even primary-level education and just over half have emotional disorders. Hossain *et al.* (2019: 1414) note that "poor access to health services, a shortage of food, and inadequate shelter are the contemporary challenges".

Among several survey-based studies, Riley *et al.* (2020: 10) find that refugees reported "horrific experiences in Myanmar, including 98.6 per cent exposed to frequent gunfire, 97.8 per cent witnessed the destruction/burning of villages, 91.8 per cent saw dead bodies, and 90.4 per cent witnessed physical violence against others". As with similar research, the authors conclude that agencies working with Rohingya should consider the impact that these experiences have on poor mental health and that "this research has the potential to inform interventions targeting such elements" (Riley *et al.* 2020: 1). But one wonders if the more useful "interventions" should be directed at preventing the ongoing violation of Rohingya human rights in Myanmar and providing more adequate housing across the border.

Photographs of refugee camps can only begin to convey the appalling living conditions faced by these refugees. The picture of one (Figure 5.3) depicts the cramped and dangerous conditions; indeed, this particular camp burned down in March 2021, leading to yet further displacement. The Begum family were forced from Myanmar to Bangladesh in 1991, and 23-year-old Anuara has never seen the former family home. She told a reporter that her camp "is like an open prison. I have spent my entire life as a refugee and still I can't find peace. My family can't sleep properly through the night because of our fear after the fire" (Ahmed 2022).

Figure 5.3 Balukhali Rohingya refugee camp in Ukhia, Cox's Bazar, Bangladesh, February 2019
Source: iStock.

The health of those living in refugee camps

One of the most well-publicized informal refugee camps in Europe was that outside Calais in northern France, known as the "Calais Jungle" (Figure 5.4), which "housed" at one time over 8,000 people between 2015 and 2016 who were seeking to cross the English Channel or were waiting for the French authorities to grant asylum. These people had made their way to France from homes in (primarily) Iraq, Iran, Afghanistan, Eritrea, Sudan and Syria; the main routes had been either via Libya and Italy, or via Turkey and Germany. An editorial in the *British Medical Journal* (cited by Davies, Isakjee and Dhesi 2017: 1264) referred to the camp as "an emblem for mass suffering of refugees". The camp was demolished by the French government in 2016, with people dispersing to different parts of France or (like many others since 2016) continuing attempts to cross the Channel in dangerous conditions and unsuitable boats (Box 5.2). Davies, Isakjee and Dhesi (2017) point out that, despite many opportunities to assist those living in the camp, the standard response has been the exercise of power through *in*action rather than the provision of tangible help. This is, the authors assert, another form of structural violence, although there is plenty of evidence of physical violence too. Structural violence is manifested in the form of wholly inadequate facilities for sanitation and inadequate supplies of food and water.

Figure 5.4 The "Calais Jungle", October 2016
Source: https://commons.wikimedia.org/wiki/File:Overview_of_Calais_Jungle.jpg.

Quantitative survey evidence of over 400 residents in late 2015 (Bouhenia *et al.* 2017) reveals that two-thirds had been the victims of assault during their journey to, or in, Calais. A similar proportion reported respiratory problems, some of which is surely due to the use of tear gas on site.

I discussed above (pp. 46–50) the increasing relevance of non-communicable diseases in resource-poor settings. This is of particular concern in refugee camps and other humanitarian settings. The Covid-19 pandemic, along with population displacement and refugee crises, has had major impacts on health systems' abilities to manage NCDs. Food supply and nutrition are greatly challenged, as are attempts to control the availability of tobacco. It is difficult to adopt healthy lifestyles in such places, and "we need to learn more about how the commercial beverage and ultraprocessed food sectors are taking advantage of these contexts" (Patel, Kiapi & Gómez 2022: 2). Allen *et al.* (2022) illustrate this in detail in conflict-ridden Libya, where the absence of stable government has allowed junk food and tobacco to penetrate in a wholly unregulated way.

Care needs to be taken not to pathologize the impacts of all those who live in refugee camps. Research by Veronese *et al.* (2020), based on interviews with Palestinian children from a refugee camp, is a good example. Although being outdoors "reminds the children that an oppressive power is present, colonizing the external landscapes, and affecting the children's internal emotional states and representations", the children's narratives reveal how they "mark out spaces of freedom and normality in temporarily liberated zones"

(Veronese *et al.* 2020: 7). Particular spaces – including schools, community play areas and a mosque, as well as the home – offer scope to maintain their wellbeing and provide "embodied resistance".

There is a more general point to be made here, expressed in an important paper by Herrick (2017b: 538), who argues that "we must also consider those spaces where pleasure and suffering intersect in ways that challenge the humanitarian impulse and crisis-led readings of health". An understandable focus on the structural determinants of poor health runs the risk of underplaying the human agency of people living in challenging conditions such as refugee camps, and their ability to resist some of the worst of such conditions. These issues have been discussed at length by anthropologists, some of whom have concentrated on "narratives of victimhood", giving voice to those experiencing harsh conditions and their domination by more powerful interest groups. Others have contested this, seeking a more positive "anthropology of the good" that emphasizes wellbeing, care, happiness, resilience and hope, albeit in challenging circumstances (see Ortner 2016 for a full account of such debates).

Refugee health in places of safety

If refugees are able to find relative safety in a country in the Global North, what do we know of their experiences and, in particular, their health status? Grove and Zwi (2006) note the long-standing othering of refugees, who are stigmatized as the deviant "them" rather than the normal "us"; less attention may be paid to the protection and wellbeing *of* the refugee and more to the protection of the host population *from* the refugee (Gatrell 2011: 98). Although some agencies have always sought a compassionate approach to saving lives and providing refuge, a more critical understanding of humanitarianism argues that it is more about "keeping strangers distant even in cases of close proximity" (in refugee camps) by policing and securing borders in order to maintain national sovereignty (Pallister-Wilkins 2020: 994). Some lives, it seems, matter more than others.

There are many stories – brief biographies – of those who have sought refuge. One such story is told in some detail in Box 5.5.

BOX 5.5 ONE WOMAN'S SEARCH FOR SAFETY: ZARLASHT HALAIMZAI

Zarlasht Halaimzai is the co-founder and chief executive of Refugee Trauma Initiative (now known as Amna: www.amna.org). Born in Afghanistan, she left Kabul – aged ten – with her family in 1992 to seek safety further north, travelling along a heavily mined road before later heading into Uzbekistan. "The next four years were a blur of trains, towns and cities, people opening their doors to us [ten family members] when we had nowhere else to go and [other] people scowling with hostility." She and her family arrived in London to seek asylum in 1996. "To survive the journey, we needed stories of hope. For us, that story was safety in London, but the reality was very different."

The bureaucracy of claiming asylum was challenging, to say the least. Forms were unhelpful (and threatening) if English language was limited. "PLEASE WRITE IN BLOCK CAPITALS." "STAY INSIDE THE BOXES OR YOUR APPLICATION MAY BE REJECTED." Proof that your life in Afghanistan had been in danger at a specific moment was required. New neighbours were often distrustful, and Zarlasht and her family experienced overt racism and personal violence. "It's difficult to describe the feeling of dislocation. People who are born in places that protect them from the misery of displacement find it hard to understand. Seeking asylum makes you feel self-conscious about your very existence. There is a feeling that pervades all your interactions, as if you constantly need to justify your presence."

Zarlasht's charity is based in Greece, where the asylum system is also "deliberately designed to be hostile". Some people resort to self-harming. Many remember, as does Zarlasht, the safety back home before the Taliban took over. Now, as she says, "history is repeating itself as they issue the same barbaric decrees against the people, confining women and children to their homes".

Source: All quotes are taken from Halaimzai (2021).

Although they are told in much less detail than that in Box 5.5, two studies of Somali refugees are instructive here, one of them in the United Kingdom, the other in the United States. Allport *et al*. (2019) asked Somali mothers living in Bristol to compare their childhood experiences in Somalia with those of their own children now living in the United Kingdom. When living in Somalia, child play was described as spontaneous and not in need of supervision, while living in tower blocks in Bristol constrained such activities; more adult supervision was required. Family and social support that was present in Somalia was absent in the new home environments. Of course, the authors note with reference to these new constraints that "the UK provided sanctuary and safety from war, with access to an education system which they observed encouraging children to develop creativity and imagination, and to health and welfare systems providing services free of charge" (Allport *et al*. 2019: 198).

Gillespie *et al*. (2020) document the mental health impacts on Somali refugees of living in North American cities, revealing the uncertainties linked to residential mobility and pointing out that housing stability is key to "ontological security" (Box 5.6). Relocating to a new country can involve multiple moves, and the authors' findings suggest evidence of discrimination and violence towards refugees, leading to symptoms of psychological distress among those forced to move but not among those moving voluntarily. These findings are echoed by earlier research on Somali refugees (Gatrell 2011: 101).

In the Global North there is considerable evidence of a "healthy immigrant" effect, meaning that immigrants tend to be healthier than the native-born population (Newbold 2018). Yet these are people who have not been forcibly displaced from their homes; refugees invariably have poorer health than immigrants as a whole, reflecting their experiences during the journey to safety and in refugee camps, and their precarious economic circumstances.

In acknowledging the constraints faced by those seeking refuge it is important to reveal the stories of those made to feel welcome. Khaled Muha and his family, refugees from Syria who spent five years in a camp in an Iraq, travelled 3,000 miles to Wales, where they have been made welcome. "A new life for us, with good people," he says. "We need to stay in Wales. Welsh people are very quiet and friendly and we are very comfortable. For me, I think here is a good place" (Crowter 2022).

BOX 5.6 ONTOLOGICAL SECURITY AND THE EMOTIONAL HEALTH OF REFUGEES

The term "ontological security" refers to the meanings people derive from feeling that they have a stable and secure life. Disruptions to stability and continuity – feelings of uncertainty – cause ontological *insecurity*.

In important, yet unpublished, doctoral research undertaken in Tasmania, Australia, Hutchinson (2010) explored how the health of refugee women from Ethiopia, Sierra Leone and Sudan has been affected by their forced separation from family members remaining in Africa. She refers to these women as being "stretched", in three senses: in terms of their physical distance from home; their emotional separation; and resulting from economic precarity. The stresses, whether physical, social or emotional, that are experienced by such women create ontological insecurity. Crucially, she deviates from a medicalization of mental ill health ("post traumatic health disorder" and its associated quantification) to argue that emotional distress is caused by these sorts of "stretchings". For Hutchinson (2010: 29), "the expression of distress by the participants in the research is the expression of social disorder and must be addressed by social remedies". An interesting example of such a "remedy" comes in learning to drive a car: this helps to access employment, and therefore some degree of financial stability as well as the ability to connect socially with others.

What comes out most strongly in Hutchinson's research is the strength of family ties and the emotional distress that arises from the severing of these. Two examples from her interviews (all quotes from Hutchinson's thesis) express ontological insecurity in very telling ways:

> People they are friendly. They are OK. They encourage you but still inside you there is something you miss. You start missing the family, ones that have been

left behind, who I left, especially my children, and I start feeling guilty or selfish. It's not OK after six weeks. What I have been through it was not good for me.
[…]
But without family it is not good for me, for me I am not happy, you see, but sometimes even when I go to school I do not understand what my teacher says. I just look at the board and don't understand because I'm missing my husband, I'm missing my family, I don't know where my family is, I don't know.

Poignantly, one woman expresses her feelings in this way: "My heart is one half in Africa, one half here."

The health of migrant domestic workers

There is a wealth of literature on the migration overseas of health professionals, doctors, nurses and other staff (see, for example, Connell & Walton-Roberts 2016 and Ormond & Toyota 2018). Here, I concentrate on the health and wellbeing of those who have moved from, or within, the Global South specifically to seek employment in other sectors, notably domestic service, with the following section concentrating explicitly on those forced to engage in sex work. Sustainable Development Goal 8 commits to "protect labour rights and promote safe and secure working environments for all workers, including migrant workers", emphasizing in particular the protection against violence and exploitation.

Migrant domestic workers (MDWs), of whom there may be as many as 70 million worldwide, are often marginalized, whether in terms of access to health and medical care, poor remuneration and treatment at work or racial discrimination; all of these contribute to poor health and health inequities. Excessive demands placed on them by employers, and the limited autonomy they have, contribute to stress. Fernandez (2018) has explored some of the issues faced by Ethiopian women employed in Lebanon. Legally, all employers are required to purchase health insurance for MDWs. However, pregnant MDWs report difficulties in accessing reproductive healthcare. Some women suffering from serious illness reported experiences of prejudice and being treated as undeserving of hospital care. The inequities faced by Ethiopian MDWs are overlain on a platform of wider inequities in the Lebanese healthcare system.

The number of MDWs in Singapore has grown tenfold in the last 35 years, now approaching a quarter of a million, and the health of almost 200 women MDWs has been explored by Anjara *et al.* (2017). Over half the women reported feeling stressed, and one-fifth felt isolated, but older women had formed stronger social networks and reported better psychological health. Complementing that quantitative research, the same group (van Bortel *et al.* 2019) note the limited agency that women felt at work, the unequal power relationships and the insecurity of employment. In contrast, other women felt quite capable of coping and showed resilience; not all felt vulnerable. But, for all, emotional and social support mattered greatly. Other studies have suggested that such support is enabled when female MDWs share living accommodation with other such workers.

Recent geographical research using data collected from 1,388 Filipina migrant domestic workers in Macao, China (Hall *et al.* 2021), has revealed associations between the density of fast food establishments and body mass index (BMI). Those MDWs living closer to Western fast food restaurants were at higher risk of being overweight or obese. Two-thirds of Filipina MDWs were overweight or obese, compared with only 6 per cent of adult women living in the Philippines. The authors suggest that the domestic workers may be purchasing and consuming cheaper food in order to save money, which they prefer to remit to family members back home in the Philippines. The general finding of links between proximity to fast food and being obese or overweight echoes much of the geographical literature on obesity and excessive weight in countries of the Global North, which suggests that minority groups and those on low incomes are similarly at risk.

Slave trades and sex trafficking

The UN defines "human trafficking" as the recruitment, transport, harbouring or receipt of persons by means of threat or use of force, with people exploited for sex or labour, and it suggests that up to a million people are trafficked each year. The very word "traffic" connotes the objectification of human beings (Gatrell 2011: 95–6). As with other forms of forced migration, there are health risks faced before, during and after the moves.

Referring back to the Sustainable Development Goals, Target 8.7 is to "[t]ake immediate and effective measures to eradicate forced labour, end modern slavery and human trafficking and secure the prohibition and elimination of the worst forms of child labour, including recruitment and use of child soldiers, and by 2025 end child labour in all its forms". To assess the extent to which this target is likely to be met, the International Organization for Migration (IOM) collects data in its Victim of Trafficking Database (VoTD). These data suggest that half report being trafficked for the purpose of sexual exploitation and almost 40 per cent refer to forced labour (mostly for domestic work). Stöckl *et al.* (2021) have used this database to document the prevalence of violence among a sample of about 10,000 people who had been trafficked. Abuse occurs throughout the sequence of trafficking, notably at the destination but also at the place of origin and during transit. The authors conclude their survey with the comment that trafficking "requires responses that are well-developed based on *individuals' different experiences*" (Stöckl *et al.* 2021: 6, emphasis added). These "experiences" are missing from their paper – and, of course, trafficking itself is, by definition, "violence". By way of contrast, the United Nations Office on Drugs and Crime (UNODC) website does provide such knowledge.

For example, Francisca Awah reports having experienced years of sexual exploitation in Europe before returning to Cameroon and being tricked into a promise of work teaching English abroad. Instead, she was exploited and abused as a domestic servant. As she tells us: "I almost sold my kidney to raise money to escape my abuser" (UNODC 2021a). On eventually returning she found herself stigmatized. Colombian Marcela Loaiza was trafficked for sex and needed money to pay for her daughter's medical treatment. Now back home in Colombia and working to raise awareness about human trafficking, she writes: "being a survivor of trafficking is like having a tattoo on the soul: no-one can see it but it is always there and remains forever" (UNODC 2021b).

Other stories relate to women trafficked from Nigeria to Europe for sexual exploitation. One young woman who was brought to Copenhagen in Denmark was taken to the city's red-light district and told that this was her new place of work.

> I looked round to see if she was pointing at a building I hadn't noticed. But no – she meant where we'd been walking. That's when she told me I was going to be a prostitute, and this was where I'd be hunting for customers. Then the whole of Denmark just crashed down on me … (BBC News 2021d)

In an attempt to offer some kind of safe space, a Danish NGO provides a van that sex workers can use. "Throughout the night, a steady stream of women arrive – men in tow – to use the van's facilities, while the volunteers stand a respectful distance away but close enough to hear if a woman is in trouble. It may be used up to 28 times in a four-hour shift" (BBC News 2021d).

Gezie *et al.* (2021) have conducted interviews with those involved in trafficking – not only victims – in Ethiopia. The factors involved in such movements operate at different levels: individual, family and community, as well as in broader structural contexts. Among the latter, poverty and female disempowerment feature strongly. The narratives of those who had returned home are telling. Relations between smugglers and the smuggled are sometimes civil at first but people are then moved on to different smugglers, with some victims in Yemen reporting that smugglers there threatened to remove body organs if money was not paid. One man told the authors:

> All of us were ordered to give the telephone numbers of our relatives who lived in countries other than Ethiopia, mostly in Saudi so they [our relatives] could send huge amounts of money to their account. The first person was severely tortured, and we were all terrorized by what they did … Most of us paid money [ranging from $500 – $3000]. (Gezie *et al.* 2021: 9)

As I have noted elsewhere (Gatrell 2011: 89), the determinants of child sex trafficking are not so dissimilar from those of adult sex trafficking across international borders, although children (like adults) are also trafficked within countries; for example, in

Thailand they may be moved among tourist destinations in order that those controlling them can avoid detection. Potential risk factors for transnational sex trafficking include poverty, low levels of education and lax law enforcement. As Merdian *et al.* (2019) note, the UK campaign group ECPAT (Box 5.7) has helped to reduce child vulnerability and has influenced the tourism industry to play its part in reducing the prevalence of such trafficking. Nonetheless, the children involved risk infection, substance abuse, violence, mental ill health and pregnancy.

BOX 5.7 EVERY CHILD PROTECTED?

ECPAT UK (Every Child Protected Against Trafficking UK: www.ecpat.org) is a charity that campaigns and advocates for the rights of children to be protected from exploitation. The organization tells compelling stories about children who have been trafficked across different countries and eventually into the United Kingdom. Here is one such story:

> Diệp was born in a rural village in Việt Nam (Vietnam). She had a difficult home life growing up following the death of the mother, caring for her father who became abusive when drinking. When she was 14, a man arrived to speak to her father and she was told she would be leaving the village for a job in the UK so she could send money home. Diệp survived a journey which took almost a year, she was made to work in a factory in Russia for several months, was sexually assaulted and in the last few months in Europe, she was made to clean different houses all day. Once she arrived in the UK, she was taken to a house and sexually exploited. She was finally identified as a child victim of trafficking a month before turning 16. Diệp was placed in care, started going to college and learning English. She was supported to claim asylum as she was terrified of the gangs finding her if she went back home. As an extremely talented artist, she was doing very well in her courses and aspired to go on to University for her degree. Her asylum claim was refused, and she spent months on end locked up in her room waiting for news about her appeal. Her mental health significantly declined; she lost all hope for the future. Diệp is 23 years old now, and she's still in limbo, never able to pursue her dreams.   (ECPAT 2022)

ECPAT (UK) is also seeking to hold the government to account. Data from the United Kingdom's Home Office reveal that in 2019/20 only 17 out of 754 children who had been trafficked into the country had been granted leave to remain – as required by international law.

Although Thailand and other countries in southeast Asia, notably Cambodia, have for many years been preferred destinations for those seeking sex with minors, other countries, such as Kenya and the Philippines, are "areas of concern" (Merdian *et al.* 2019). As one of the participants in the authors' roundtable put it: "The offenders are rich, they can buy people off. They can buy the family off, they can groom the family. They can pay police officers off. Families will take any amount of money for a child. So poverty, corruption and richness together is the perfect triangle of abuse" (quoted in Merdian *et al.* 2019). ECPAT reports on the progress each country is making towards dealing with transnational trafficking.

Of course, much of the "traffic" involving young people, as well as adults, is conducted online. Those involved are not on the move but are trafficked virtually across a variety of networks and platforms.

Natural hazards and the health of those displaced

Natural hazards, such as cyclones, tsunamis and earthquakes, force people from homes. Aside from the direct effects of flooding, drowning and falling debris, the disruption of food and water supplies also causes illness and mortality. In their systematic review of the health impacts of such population displacement, Jang, Ekyalongo and Kim (2021) suggest that, globally, in 2018 over 17 million people were newly displaced because of natural disasters. In east Asia and the Pacific region and south Asia alone, more than 12 million were internally displaced. The economic losses (to fishing industries and agriculture, for example, as well as reduced tourism) incurred as a result of such disasters have led to increased poverty, and, unsurprisingly, studies report increased levels of anxiety and depression among both adults and children displaced (Jang, Ekyalongo & Kim 2021).

The 26 December 2004 tsunami that resulted from an earthquake off the Indonesian coast is thought to have killed almost 250,000 people (of whom about 170,000 were in Indonesia), but it also caused massive population displacement, with clear consequences for mental health. Frankenberg *et al.* (2008) have assessed the impact of the tsunami on those living on the Indonesian island of Sumatra, by comparing survey data collected before and after the disaster; the total sample size was over 20,000. Using the "post-traumatic stress disorder" checklist, the authors find in their statistical analysis that "proximity to the coastline (a good proxy for experiencing the tsunami), exposure to traumatic events, loss of family and friends, and loss or damage of property were strongly related to post traumatic stress response" (Frankenberg *et al.* 2008: 1675). Whether this entirely predictable finding has improved the lot of those living in the devastated regions of western Sumatra is a matter of debate.

Sheller (2013) draws on a mobilities perspective to show how, following the earthquake in Haiti in 2010, many of the poorest were left isolated and disconnected ("marooned on an island of misery", in her memorable phrase: 187) and living, months later, in tents and informal shelters. Such conditions exposed people to danger, as well as poor water quality and sanitation, and contrasted with those experienced by aid workers living in good accommodation. The destroyed buildings and heaps of rubble were a "constant traumatizing reminder of death" (Sheller 2013: 188); around 200,000 are thought to have died. Some Haitians moved abroad after the earthquake; Montreal, in French-speaking Canada, was one destination. Gautier *et al*. (2020) interviewed 23 people who had relocated, documenting financial worries and problems finding work, as well as discrimination when seeking employment. Chapter 7 discusses further the consequences of the Haitian earthquake, notably the controversial involvement of UN peacekeepers and the role they played in the subsequent outbreak of cholera.

Concluding remarks

There are many associations between population movement and health and wellbeing. My emphasis here has been on the health and wellbeing of those who are marginalized: those who move not by choice but as a result of war, conflict, environmental disaster or government projects. Such concerns add an important dimension to the growing attention in social science (and particularly geography) to mobilities (see, for example, Adey *et al*. 2015). Much of the earlier literature focused on global elites and their movement enabled by air transport or automobile; in turn, it had little to say about the health impacts of such (Gatrell 2011). But the increasing flows of people who have little choice about whether, how and where to move create impacts on their health and wellbeing that deserve the attention of geographers. Survey research can help, but the narratives offered by those moving are invariably more telling.

There is another dimension to such forced population movement – that driven by climate change – and this is discussed further in Chapter 9. In addition, some movements in the Global South are associated with infectious disease, and these are considered in Chapter 8.

All these flows take part in the wider context of globalization and a set of networks that include not only the human actors themselves but a set of other actors and materials, including borders, documentation, camps and hazards of various sorts – whether natural ones or those created by people (or governments) who do not have their interests at heart.

Further reading

Boyle (2021: ch. 11) provides an excellent introduction to migration, including issues of securitization. My earlier review of literature on the health of the displaced and dispossessed (Gatrell 2011: ch. 6) may be worth consulting. For a comprehensive overview of refugee and immigrant health, including vignettes on 30 countries, see Kemp and Rasbridge (2004). An important series of brief papers on borders, refugees and mobilities is in Minca *et al.* (2022), but see also Pallister-Wilkins (2020, 2022). Weber and Pickering (2011) discuss the deaths resulting from those attempting to cross borders.

The UN has produced a "Global Compact for Migration", including refugees (see UN Refugees & Migrants 2018), but this has been criticized by some writers, such as Pécoud (2021). Ferris and Donato (2020) discuss in detail the issues of global governance involved in the international movements of refugees, and this is worth consulting in light of the discussions in Chapter 4.

Amzat and Razum (2022: ch. 9) deal with migrant health, and complement some of what is written here. I have said nothing here about health or medical "tourism" (the movements abroad of those seeking healthcare or medical interventions) but do address that subject briefly in Gatrell (2011: 174–8). For a more recent overview, see Ormond and Toyota (2018).

6
MATERIALS ON THE MOVE: OUT OF THE GROUND, AND ACROSS THE GLOBE

> [A]n urgent need exists for multidisciplinary health research that describes in greater, context-specific detail the direct and indirect health effects of what might be called a global extractive order.
>
> Schrecker, Birn and Aguilera (2018: 142)

My focus here is on the extraction and movement of materials that are either health-damaging or whose transfer across geographic space results in potential harm. To that end, I concentrate on: mineral extraction; the transfer of waste, especially plastics; and arms transfers. These are linkages or commodity chains that all involve movement, including the extraction of materials from the ground, the transport of these and, ultimately, the disposal of waste products associated with extraction or with the residue of consumer products. Inevitably, those living near extraction or waste sites, or living alongside the routes along which materials are transported, incur environmental penalties or injustice (Box 6.1). Although many of these transfers can result in harm, in some cases there might be benefits; for example, the extraction of precious minerals may offer employment to those engaged in extraction and movement (although, as we see shortly, costs invariably outweigh benefits). Some may benefit, but others reap no rewards and incur harm.

BOX 6.1 GLOBAL ENVIRONMENTAL JUSTICE

There is plenty of evidence in the Global North that people living in areas of socio-economic deprivation, or areas with high concentrations of people of colour, suffer inequitably both from the location of hazardous or noxious facilities in their neighbourhood and from lacking any involvement in decision-making about such locations (Crighton, Gordon & Barakat-Haddad 2018). Risks are unevenly distributed, socially and spatially, and such environmental injustice carries over into many countries of the Global South. Further, as Walker (2009) has observed, the scope of environmental justice now embraces issues that cut across international boundaries, involving relations between countries and global issues such as waste transfers and climate change, both of which are considered here and in Chapter 9.

Indigenous communities across the globe suffer disproportionately. Crighton, Gordon and Barakat-Haddad (2018: 41–2) remind us that the people of the Aamjiwnaang First Nations in Ontario have reported respiratory illness, reproductive problems and elevated cancer rates from being exposed to air pollution from chemical plants and oil refineries in the province's so-called "Chemical Valley", and that a creek running through their reserve has been shown to be contaminated with arsenic, cadmium, lead and nickel.

See Environmental Justice Atlas (https://ejatlas.org) for an interactive atlas of global environmental justice that highlights specific issues of environmental and health concern. For discussion of a similar map of environmental injustices in Brazil, as well as an excellent discussion of links to political ecology, see da Rocha *et al.* (2018). Importantly, some conceptions of environmental injustice do not give sufficient attention to "the broader postcolonial terrain of plural injustices and violence" (Akese & Little 2018: 77). There are layers of toxicity that include not just the environmental contaminants but also the poverty, hunger and lack of care that form the backcloth to resulting ill health (see Swartz *et al.* 2018 for a fine example of "toxic layering").

Mineral extraction and global health

In Chapter 2 I explored some dimensions of health inequalities in Mozambique. Others are reflected in mining. Among the mining projects in Mozambique is the Montepuez open-pit ruby mine in Cabo Delgado, started by the major company Gemfields in 2011. Its impact on population health (along with that of other mines in sub-Saharan Africa) has been studied by Leuenberger *et al.* (2021). In focus group discussions, people reported their lives being threatened if they attempted to scavenge the mine tailings. Others reported ill health resulting from population mixing. For example, one person complained about the importation of diseases from those coming to work in the ruby mine: "Now these people from outside who come with their diseases come here and contaminate us without knowing the reason why the diseases are increasing more and more" (quoted in Leuenberger *et al.* 2021). The authors also report health effects linked to pollution and food insecurity. Despite attempts by the company to improve living and working conditions, the overall view on health was negative.

Gemfields has extracted considerable wealth from its ruby mine. Mozambican investigative journalist Estacio Valoi suggested in 2016 that it had generated $122 million in revenue since 2014 (Valoi 2016). But the local elite has benefited too. General Raimundo Pachinuapa, a member of the ruling party Frelimo, and former governor of

Cabo Delgado province, is alleged to have appropriated land belonging to the farmer who initially discovered the rubies. Locals allege that they have been beaten and murdered, and their crops and homes burnt down by local police and private security forces. In January 2019 Gemfields paid £5.8 million (about $7.6 million) to community members residing near the Montepuez ruby mine in a "no admission of liability" move that settled a claim of human rights abuses brought against it by locals (Hunter & Lawson 2020).

Resource extraction in the Democratic Republic of the Congo has included diamonds, copper and – especially – cobalt (used to make rechargeable batteries for laptops, smartphones and electric cars: it has been described as the modern-day "oil" of a low-carbon economy); all these minerals are found extensively in Katanga province in the southeast of the country (Figure 2.8).

Such mineral extraction has serious consequences for those living near the mines. Research suggests that thousands of people are exposed to dangerous levels of toxic pollution, which has caused birth defects among the children of those working in the mines (van Brusselen *et al.* 2020). Examining the political economy of cobalt extraction, Sovacool (2019: 916) recognizes some benefits, such as increased revenue, but sets this against the many disbenefits, namely "accidents and occupational hazards, environmental pollution and degraded community health, exploitation of miners and unfair market practices, the erosion of democracy *via* corruption and malfeasance, displacement of indigenous peoples, and violent conflict and death". The informal (artisanal) miners suffer the worst. As one of Sovacool's interviewees tells it:

> The conflict between industrial and artisanal miners is very real. Because the main mining sites are now so secure, artisanal miners have been pushed to the periphery. I see them picking through the waste dumps and tailings ... [W]hen we catch them doing that, of course we beat them before we call in the mining police. (Sovacool 2019: 933)

He concludes that the extraction of cobalt for profit is part of a complex political economy whose power dynamics cement inequality. That complex political economy now includes the ownership of some mining companies by Chinese corporations, workers for which allege "discrimination and racism, reminiscent of the colonial era", as well as poor pay. As one worker put it: "We're being treated in a bad way by the Chinese. I'm a victim of assault myself. I was slapped across the face four times" (Pattisson 2021; see also the London-based "watchdog" Rights and Accountability in Development: www.raid-uk.org).

As Schrecker, Birn and Aguilera (2018) point out, and as shown in Chapter 2, such extraction and "plunder" is nothing new, whether in South America or Africa. The health consequences of extraction (poisoning and respiratory diseases, for example) result from processes operating at various scales: internationally these include global corporate power and military conflict, but more locally they involve unsafe workplace practices and forced displacement.

Not all mining is to extract precious minerals for wealthy citizens or modern industries in the Global North, nor are the health risks attributable to mining and their impacts on marginalized communities restricted to those living in the Global South, as research on Indigenous groups in the United States highlights. Lewis, Hoover and MacKenzie (2017) have reviewed numerous studies of health inequalities linked to uranium mining among Native Americans. For example, whereas nationally one-quarter of the deaths of over 4,000 former uranium miners were from lung cancer, Native miners from the Navajo tribe suffered disproportionately and the incidence of serious lung disease was four times higher than among non-Hispanic whites. Another study of Navajo births reported by Lewis, Hoover and MacKenzie (2017) reveals an excess of birth defects and other adverse birth outcomes, linked to mothers living close to uranium mines or the associated mine dumps.

The contribution of coal to global energy consumption – and, of course, to emissions of CO_2 and, therefore, global warming – is well known. But coal production generates vast amounts of waste material: rocks, ash, slags and acid mine drainage; these release metal-rich effluent into soils and water courses. A spatial analysis of soils in the vicinity of coal-mining areas by Xiao et al. (2020) suggests that the highest concentrations of potentially toxic elements are to be found in southeast Asia, south Europe and north Africa. Nickel, cadmium, zinc and mercury (and arsenic, to an extent) are examples of such elements, conferring risks of both cancer and other diseases when they are taken up by plants, including wheat and vegetables.

Among the wealth of epidemiological evidence, consider the following studies outside the Global North. Liao et al. (2010) note that extensive coal mining (25 per cent of national output) in Shanxi province in northern China is associated with the highest rate of neural tube defects (such as spina bifida) in the country as a whole (106 per 10,000 births in 1987, and 61 per 10,000 births in 1996–2002). Although some of this may be attributed to daily passive exposure to cigarette smoke and poor ventilation during heating, it is telling that the rate of birth defects within 6 km of coal mines was 271 per 10,000 births, compared with 202 per 10,000 births further away.

In India, Mishra (2015) conducted 300 interviews in an industrial part of Odisha state (formerly known as Orissa). Interviews were conducted in seven villages located at varying distances from coal mines. Ninety per cent of households living near the mines reported health problems (headache, skin diseases, arthritis), compared with only 50 per cent further away. Mishra attributes these to polluted drinking water that is sourced from the opencast mines.

Environmental and epidemiological studies such as these are valuable. But Morrice and Colagiuri (2013) refer to the differential power relationships between the coal-mining industry and local communities whose health may be impacted. The former has more financial clout, generating social and environmental injustice, since the most disadvantaged communities suffer the ill health that results from living in or near hazardous environments. As the authors say, such power differentials are another determinant of health inequalities, in the Global North and the Global South alike.

Similar studies using environmental monitoring and epidemiological investigations have been undertaken by those looking at oil fields. As with coal, in the extraction process billions of gallons of untreated waste and crude oil are released into the environment. In the Peruvian Amazon contaminated water resulting from oil extraction has serious health consequences. O'Callaghan-Gordo *et al.* (2021) sampled blood lead levels in over 1,000 indigenous inhabitants (both children and adults) living close to oil production sites, finding elevated levels (potentially leading to a range of health problems – neurological, cardiovascular and renal, among others) in adults and children consuming fish. Related research provides evidence that lead has entered the food chain, with high levels in tissues of wildlife species frequently consumed by Indigenous peoples. Higher rates of lead levels in blood were also linked to the use of river water for drinking and bathing. The rivers are major tributaries of the river Marañón, which is itself a tributary of the Amazon.

Whether such exposure leads to elevated risks of cancer is debated in the epidemiological literature. For example, a small-scale study in the village of San Carlos, Ecuador, finds that residents were exposed to potentially carcinogenic petroleum hydrocarbons; a significant excess of male cancers were found (San Sebastián *et al.* 2001). This was supported by other research in the Ecuadorian Amazon basin (Hurtig & San Sebastián 2002). Comparing cancer rates for those living in large areas with a history of oil exploration and areas without oil development suggests that there were significantly higher rates in exposed areas. Conversely, and also working in Ecuador, Moolgavkar *et al.* (2014) find that there were no significant differences in deaths from cancer between those areas with and without oil exploration and production. It should be noted that this study was supported financially by the Chevron Corporation.

In Iraq, oil extraction, coupled with poverty and the consequences of war, has had devastating consequences for those living close to oil production facilities. Flare stacks burn the by-products of oil extraction and release chemicals (such as benzene, black carbon and carbon monoxide) that have been shown to cause a variety of serious health problems. Detailed interviews reveal these in stark terms. For example, 60-year-old Kamila Rashid points a finger at the nearby refineries run by the Kurdistan energy company KAR. "KAR uses the water from the river for their work and then they dump their waste," she says, pointing at a pipe piercing through the exterior wall of the refinery. The waste flows directly into her family's field, she says, and KAR also dumps its waste in the nearby river. The crops no longer grow, Rashid says, and, anyway, the land "isn't safe for agriculture now. And it isn't safe for us" (quoted in Billing 2021). Further south in the country there are major environmental and health effects resulting from the US-led invasion of Iraq in 2003 (and the earlier military operation to remove Iraqi troops from neighbouring Kuwait). Areas surrounding Basrah were exposed to by-products of the petroleum fires that occurred during both conflicts, and the region was exposed to chemical weapons agents, depleted uranium and benzene. Hagopian *et al.* (2010) report almost 700 cases of childhood leukaemia in and near Basrah between 1993 and 2007, with dramatically increasing rates between 1993 and 2007 (2.6 per 100,000 in 1993; 12.2 in 2006), substantially higher than Iraq's neighbour Kuwait and countries in the Global North. In Iraq as a whole the invasion had other consequences. US military bases used burn pits to dispose of military and industrial waste – paint, plastics, used medical supplies, electronic equipment, spent weapons, rubber, and so on. Incineration of this waste produces particulate matter and dioxins, which are well-known carcinogens. But the health consequences have migrated west, with various studies revealing severe health complications for American veterans (Coughlin & Szema 2019).

Aside from the pollution resulting from the extraction process, the transport of oil can lead to massive oil spillages, with consequences for both marine life and subsequent human ill health. But the movement of oil can itself cause loss of life. Carlson *et al.* (2015) have identified 28 major incidents in sub-Saharan Africa, the majority of which were in Nigeria. Most of these were the result of scavenging fuel leaking from damaged pipelines or tankers. In January 2019 an overturned oil tanker exploded in Odukpani in Cross River state and up to 60 people were killed when scavenging the leaking fuel (Guardian 2019). An explosion at an illegal oil refinery killed over 100 people in Imo state in Nigeria in April 2022 (BBC News 2022b). There is an element of victim-blaming, since these "accidents" invariably result from poor maintenance and inadequate safety measures. Elsewhere in Nigeria journalists suggest that "environmental destruction by multinational oil and gas companies has left a profound legacy, with pollution evidence in the creeks, rivers and air" (Akinwotu 2022).

The last words in this section must go to Schrecker, Burn and Aguilera (2018: 137). "Accelerating in the postwar consumer era, market-and profit-driven growth – accompanied by militarism and magnified by neoliberal globalization since the 1980s – has generated ever-growing demand for all sorts of goods, with little regard for the health, environmental, and resource depletion consequences". The same authors point a very critical finger at the World Bank for its role in sustaining environmental and health damage, or what they call "extractive injustice".

Waste assemblages

Waste is "matter-out-of-place" (Van Loon 2002: 106) and it has both real and potential consequences for human health, as well as the health of the environment. The impacts on air and water pollution are covered in the following chapter. My focus here is more on the plastic and electronic waste products from industry and how they move from centres of production to be disposed of, mainly in the Global South. Indeed, the hazardous wastes that flow from north to south are the mirror opposite of the flows of raw materials (including those considered above) from south to north. Kirby and Lora-Wainwright (2015: 40) suggest that waste imports (notably from the United States and Japan) have assisted China's "economic miracle", noting that "even the most cursory exploration of China's sprawling cities and vast hinterland reveals abundant evidence of intensive formal and informal [waste] resource extraction, particularly in the nation's coastal zones, devouring everything from foreign automobile scrap to coat hangers".

Concerns over the international transport of hazardous wastes led in 1992 to ratification of the Basel Convention, which sought to reduce waste volume and to restrict such movement "except where it is perceived to be in accordance with the principles of environmentally sound management" (Secretariat of the Basel Convention n.d.). Whether the convention has had any real impact on waste generation and transfer is a moot point. As Walker (2009) suggests, there are inequalities in waste transfers, processing and disposal, and he criticizes failures of international negotiations and agreements, including the Basel Convention. Kirby and Lora-Wainwright (2015: 41) refer to the treaty as "a frequently vague and contradictory document riddled with assorted loopholes and exceptions".

Waste companies have been shown to be cavalier in what they ship for disposal overseas. For example, in 2021 the United Kingdom's largest waste company, Biffa (annual turnover in excess of £1 billion), was fined £1.5 million after exporting more than 1,000 tonnes of household waste marked as "waste paper" for recycling in India and Indonesia, when the material included metal, condoms and clothing (Taylor 2021).

If we wish to think, with Van Loon (2002), in terms of assemblages (see above, pp. 8–10), it is clear that the raw materials, minerals, labourers, transnational companies and waste products form a network or nexus that may sustain parts of the global economy but potentially do great harm to those who lie on the periphery of that network. Indeed, Van Loon (2002: 108) suggests that "one could argue that the world risk society [p. 11 above] is above all a world waste society", which is enabled by pollution havens (Box 6.2).

BOX 6.2 POLLUTION HAVENS

If countries have adopted stringent environmental regulations that restrict the domestic recycling of hazardous waste products, opportunities exist for them to export to countries where such regulations are weaker. For example, used batteries in the United States get sent to Mexico, where the lead is extracted. The number of such batteries shipped to

Mexico grew by 400 per cent between 2005 and 2010 because of stricter regulation in the United States. Mexico therefore becomes a pollution "haven". The health consequences may be severe. For example, workers in a car battery recycling plant in Monterrey have been diagnosed with acute lead poisoning. "Elevated levels of lead have turned up in workers' blood, test results show, as well as in nearby lots where children play and families raise livestock, according to interviews with residents and current and former factory employees" (Partlow & Warrick 2016).

More generally, there are risks that countries in the Global South attract foreign investment because of low labour costs and lower environmental standards. Clapp (2002: 11) puts this dramatically: "The intensity of dirty industry is rising in the developing world just as it is falling in the industrialized world."

In contrast to a pollution haven, there is the (controversial) concept of a pollution "halo", which suggests that multinational companies use green technologies when they are investing in host countries.

The focus here is more on the consequences for human health, particularly for "waste pickers" living in or near landfill sites in middle- and low-income countries. Shibata *et al.* (2015) undertook a study in Makassar, Indonesia, interviewing 113 waste pickers and documenting health outcomes and household characteristics, as well as measuring air quality and waste characteristics. They find that children suffered from diarrhoea and respiratory infections, as well as being at risk of injury from bulldozers, hydraulic shovels and traffic dealing with the waste; they were also exposed to biological and chemical hazards in the absence of any protective clothing. As the authors put it: "The waste picker and his or her family's exposure to occupational and domestic hazards, health status, and well-being is an important dimension embedded in the political economic and environmental contexts in which they live and work" (Shibata *et al.* 2015: 409). The wastes that are handled in sites such as this include plastics and electronic wastes; what follows is an overview of the health impacts of dealing with each.

Plastics on the move

The trade in plastic waste, and particularly its impact on the marine environment, has gained considerable public attention in recent years – and even more so with micro- and nanoplastics (MNPs). Primary MNPs are those added to cosmetics and detergents, with secondary MNPs resulting from the degradation of larger plastic products; both are ubiquitous in food and drinking water, and have been detected in the human placenta. Although they were not concerned with the health impacts, Zhao *et al.* (2021) studied the spatial and temporal structure of trade in plastic waste across the globe: see also Box 6.2. Imports and exports of plastic waste both increased by over 140 per cent from 2001 to 2016, to almost 11 million tonnes, with the United Kingdom and United States generally located at the core of the global network. Since 2017 (when China implemented a ban on imports of plastic waste) Thailand, India, Vietnam, Malaysia, Indonesia and Turkey have become prominent net importers, although import bans are gradually coming into force. The European Union has banned the export of plastic waste to countries in the Global South. The Covid-19 pandemic will, inevitably, have led to a major increase in the production, circulation and discarding of the plastic waste used in personal protective equipment.

Jiang *et al.* (2020) report that only about one-quarter of plastic waste is appropriately recycled and incinerated; the rest is burnt in open pits or discarded, leading to environmental pollution. These authors consider the health effects on marine and freshwater animals and conclude that it is entirely plausible that MNPs can accumulate in the human body and affect health. Yet, although MNPs are a concern to people living in the Global North, there are real risks to the health of those in the Global South from larger items of plastic waste. As we see in Chapter 8, discarded rubber tyres and old plastic containers retain standing water, which provides an attractive breeding ground for the mosquitoes that transmit disease.

E-waste on the move

Major sources of electrical and electronic waste (e-waste) include domestic washing machines and refrigerators, computers and laptops, televisions and mobile phones, along with various components including batteries, circuit boards and cathode ray tubes. The burning of e-waste (such as electrical cables) seeks to recover valuable metals, particularly aluminium and copper. The United Nations Institute for Training and Research (UNITAR) records 53.6 million metric tonnes (Mt) of e-waste in 2019, an increase of over 20 per cent since 2014. The *Global Transboundary E-Waste Flows Monitor: 2022* report (Baldé *et al.* 2022) suggests that Asian countries generated the greatest volume (24.9 Mt), followed by the Americas (13.1 Mt) and Europe (12 Mt), while African countries generated 2.9 Mt (see also IMPEL [European Union Network for the Implementation and Enforcement of Environmental Law] 2020). Only about a fifth of this waste is collected and recycled adequately, the remainder being dumped or recycled by informal workers (Krishnamoorthy *et al.* 2018). In China, Kirby and Lora-Wainwright (2015: 40) refer to the toxic "cooking" of such waste in workshops, where "hundreds of thousands of migrant labourers toil at such de-manufacturing in dangerous conditions".

The Basel Action Network (BAN) installed GPS trackers in old computer equipment to monitor the flow of e-waste, finding clear evidence of illegal shipments, the volume of which, "if extrapolated, would total 352,474 metric tonnes per annum, moving from the

EU to developing countries. This amount could fill 17,466 large-size intermodal shipping containers. If they were loaded onto trucks, the trucks would stretch back-to-back [sic] for 401 kilometers" (BAN 2019). The United Kingdom appeared to be the worst offender, with the bulk of its e-waste exported to Africa. It should be noted that BAN's mission also includes monitoring plastic waste pollution.

Dioxins and related compounds are released as by-products in e-waste recycling, and accumulate in soils and thence the food chain. Among the many health effects of dioxins are birth defects, cancer and neurological damage, and there is plenty of evidence from China of high concentrations of these toxic elements at or near e-waste sites (Dai *et al.* 2020). Other research has contrasted blood lead levels among children living in the town of Guiyu (a major centre for e-waste handling, referred to as "Treasure Town" in the paper by Kirby and Lora-Wainwright (2015) in Guangdong province, China, with those children living in the neighbouring town of Chendian. Huo *et al.* 2007 find that blood lead levels were significantly higher among children living in Guiyu (Table 6.1), where up to 80 per cent of families were involved in e-waste handling in small workshops; exposure is probably a consequence of the use of lead when dismantling cathode ray tubes in TVs and monitors, and in handling solder.

In an important paper on e-waste, Akese and Little (2018) have looked in depth at one site, Agbogbloshie, a scrap market in Accra, Ghana, that (until July 2021) processed e-waste and provided employment for up to 6,000 people (Figure 6.1). Considered by some as one of the most polluted places on earth, Greenpeace has recorded very high levels of cadmium and lead in the scrapyard's soil. Akese and Little (2018) point to the environmental injustice engendered there, but they also point out broader concerns.

> Thinking through the multiple injustices … suggests Agbogbloshie is not simply a site of e-waste contamination, but a landscape produced by the myriad ongoing forms of violence be they ecological or socioeconomic against marginalized populations in postcolonial Ghana. Harm from e-waste is only one iteration of a longer mostly invisible socioecological violence.   (Akese & Little 2018: 82)

They draw on the concept of "slow violence" to illuminate their argument (Box 6.3).

Table 6.1 Blood lead levels (in µ/dL) in a town handling e-waste (Giuyu) and a comparison town (Chendian), China

	Giuyu		*Chendian*	
Age (years)	Number	Mean blood lead level	Number	Mean blood lead level
1–4	22	12.88	14	10.02
4–5	49	14.01	20	9.46
5–6	94	16.54	27	10.27

Source: Huo *et al.* (2007).

Figure 6.1 Extracting e-waste at the Agbogbloshie scrap market, Accra, Ghana
Source: iStock.

More epidemiologically oriented research on the same site has been conducted by Nti *et al.* (2020), who point out that the open-air burning of sheathed cables alongside other waste (discarded car tyres, old refrigerators, and so on) releases a mix of harmful substances, such as particulate matter, polyaromatic hydrocarbons and heavy metals. Their research compared 142 e-waste workers with 67 control participants from Accra, and they find that lung function was markedly worse, and asthma elevated, in the former. But, as Akese and Little (2018) have pointed out, there is a wider socio-political context, and Akese has documented the Ghanaian government's sudden demolition of the site, displacing several thousand workers who had depended on it for employment and income (Akese, Beisel & Chasant 2022).

BOX 6.3 FROM STRUCTURAL VIOLENCE TO SLOW VIOLENCE

The title of this box is borrowed from Thom Davies (2022). I referred in Chapter 1 to the concept of structural violence. It was Rob Nixon (2011) who introduced the idea of slow violence as that which is different from the immediate, and direct, physical violence to which individuals may be subjected. Akese and Little (2018) highlight three aspects to it. First, it is incremental. Second, it has "ambiguous spatial and biophysical boundaries"

(Akese & Little 2018: 82). Third, it is invisible. Davies suggests that structural and slow violence have much in common. Both are representative of the inequities that we looked at in Chapter 2.

But Davies (2022) questions the "out of sight" component of slow violence, suggesting very plausibly that those exposed to toxic chemicals and other hazards see it every day. In particular, his own ethnographic work in Freetown, a small Louisiana town on the river Mississippi (located in close proximity to numerous parts of the petrochemical complex, and part of what is popularly known as "Cancer Alley"), refers to "frequent reports of skin irritations, dizziness, sinus infections, and headaches linked to chemical pollution, forming an embodied means of beholding slow violence" (Davies 2022: 424). As he indicates, people living in Freetown face a type of environmental injustice that might also be called environmental racism. Tellingly, he suggests that ignoring lay perspectives – the lived experiences of those dwelling near petrochemical industry – is a form of another kind of violence: "epistemic violence". First introduced by scholar Gayatri Chakravorty Spivak (1988), this concept means the silencing of marginalized groups, and reminds us of work undertaken by those in the Global South to offer contributions to knowledge that differ from those working in the Global North.

Arms on the move

Having spoken above about structural and slow violence in the context of waste handling, I consider here the physical violence (the gendered nature of which was considered in Chapter 2) that results from the shipments and deployment of military material, a trade that is worth almost $2,000 billion a year (Feinstein & Choonara 2020), as well as some health impacts from its use.

The Stockholm International Peace Research Institute (SIPRI) holds a database on international arms transfers and can be used to examine sender and receiver countries, disaggregated by arms type (such as aircraft, artillery, missiles, and so on). The top three exporting countries (between 2010 and 2020) were the United States, Russia and India, while the top three receiver countries were India, Saudi Arabia and China. Saudi Arabia has received light armoured vehicles from Canada, Hawk 100 aircraft ("an advanced two-set weapon system trainer with ground attack capability") from the United Kingdom and multiple supplies of aircraft, helicopters, missile systems, guided bombs and tanks from the United States. Much of this material has been used to support Saudi Arabia's interventions in Yemen (Figure 6.2), designed to prevent further incursions by the Iran-supported Houthi Shia group.

Ferguson and Jamal (2002) are critical of Canada's role in arms exports to Yemen, arguing that it fails to consider the negative impacts of such exports on people's health

Figure 6.2 "Living" in Yemen: the devastation of war
Source: iStock.

and healthcare. They note the direct link between human suffering and arms flows, with women and children accounting for the vast majority of those adversely affected. But there are of course indirect effects: the destruction of healthcare facilities, population displacement, loss of employment and shortages of food, safe drinking water and medicine.

Among these indirect threats are both environmental ones and enduring threats to human life, powerfully outlined by Griffiths (2022: 282–3), who describes the violence of war in late modernity as "subterranean, hydrological, agricultural, molecular, cellular, genetic, teratogenic and carcinogenic". He summarizes the long-term effects on health in Gaza, Afghanistan and Iraq. For example, in Fallujah and Basra, cities in Iraq, very high rates of miscarriage and birth defects are associated with high levels of toxic metals (uranium, mercury, lead) left as residues from weapons. Griffiths further notes that epidemiological studies such as these are often discounted or ignored by those in positions of power (in Washington and Geneva, for example). But even they cannot discredit evidence pointing to severe contamination of groundwater and soils. Geography matters: "Geographical approaches could identify munition residues and therefore aid clinical management, document their spatial distribution of threat to different communities and determine their provenance in terms of arms production chains, thereby opening lines of political accountability" (Griffiths 2022: 293). The impacts on children caught up in the conflict that results from arms sales are particularly distressing, with terms such as "collateral damage" serving as whitewash.

In a powerful conclusion to their paper, Feinstein and Choonara say that the arms trade is poorly regulated and riddled with corruption. "National security threats are blown out of all proportion, while climate change, health risks and other very real threats to human security are deemed secondary and consequently underfunded. The impact on child health is devastating" (Feinstein & Choonara 2020: 4).

Given this, it might seem odd to read a rather dispassionate spatial analysis of child mortality, but such a study has been undertaken, for the entire continent of Africa, by Wagner *et al.* (2018). The study suggests that between 1995 and 2015 up to 5.5 million deaths of children aged under five were related – directly or indirectly – to armed conflict; this equates to about 7 per cent of all such deaths. As the authors point out, there are longer-term risks because of the children living in what they call a "compromised environment" (Wagner *et al.* 2018: 860); in other words, the conflicts destroy the infrastructure required for survival, including water, sanitation, food, healthcare and other resources. At a global level, Hyatt (2006) has analysed data for over 80 countries to reveal an association between infant mortality (and life expectancy) and military expenditure; controlling for other factors, he shows that countries that spend more on arms are those with poorer health outcomes.

Of course, maps and statistical analyses such these do nothing to convey the reality of human suffering. However, maps do serve to highlight those regions in the continent of Africa that have seen the most casualties. Conflicts in Sierra Leone, Rwanda, Ethiopia and the Democratic Republic of the Congo, and associated child deaths, are readily apparent. But the photographic evidence, and the stories told by those living with conflict, are more compelling.

Concluding remarks

The movement of material goods from country to country and the resulting consequences for human (and environmental) health form a key part of the geographical structuring of global health. This chapter has considered the health associations with things that flow, whether oil and gas and their pollution of flowing rivers, plastic and e-waste transported across the oceans, arms traded among countries, or minerals extracted from the ground and exported for use and sale, primarily in countries of the Global North. The trade in waste tends to flow from the Global North to the Global South, producing environmental injustices that are felt regionally and locally.

Violence, whether physical, structural or slow, characterizes all these flows. With notable exceptions, health geographers have yet to attend to these linkages.

Further reading

Boyle's excellent book (2021) has a brief discussion of waste, set within a broader discussion of what he calls "humanity's war on the earth" (263). This is of relevance to the following chapter and to Chapter 9, on climate change. See also O'Neill (2019) on the mobilities of waste across the globe. The research reported by Akese and Little (2018) merits a careful reading.

For a fine discussion of links between resource extraction and human health, from a political economy perspective, see Schrecker, Burn and Aguilera (2018), who adopt a broad definition of extraction that includes water, sand mining and land grabs. Ramirez *et al.* (2017) have reviewed in detail the environmental and health impacts of oil extraction.

On the arms trade and health impacts, see Feinstein and Choonara (2020), but also an earlier paper by Mahmudi-Azer (2006). As noted in the text, Griffiths (2022) is an important recent statement on the impacts of late modern war.

7
AIRS, WATERS AND PLACES

> The world is a smaller place than it once was and other people's pollution crowds in on us.
> Yearley (2000: 147)

The previous chapter considered the impact that mineral extraction, waste mobilities and arms transfers are having on environmental quality as well as human health. These impacts are transmitted through the air and via water courses. The present chapter considers this pollution, and its consequences for human health, in much greater detail. However, although there is a truly vast literature on air pollution, water quality and their health impacts across the globe, attention here is focused primarily on how these impacts are felt unequally in different places, concentrating – as elsewhere in the book – on places in the Global South. Much of the literature on air and water pollution and health is epidemiological; although some of that research is touched on here, the emphasis is more on the political economy/ecology of air and water quality. Even though this is a generalization, it is invariably the case that those bearing the greatest burden of air pollution are also those in the poorest of households: a clear example of environmental injustice or inequity.

To give a brief indication of the scale of the problem, Fuller *et al.* (2022) suggest that pollution (of both air and water) is responsible for about 9 million deaths every year, an increase of 7 per cent since 2015 and more than 66 per cent since 2000. They write that "little real progress against pollution can be identified overall, *particularly in the low-income and middle-income countries, where pollution is most severe*" (Fuller *et al.* 2022: e535, emphasis added), and note that there are synergistic effects with climate change and loss of biodiversity. Air and water pollution are demonstrably planetary threats, with causes and impacts on human health cutting across national boundaries and therefore requiring international attention.

The political economy of pollution

The hypothesis of a pollution "haven" was introduced in the previous chapter with reference to waste, but is equally relevant in a discussion of air and water pollution. To recap, a haven effect means that firms in high-income countries seek to locate production offshore in low- and middle-income countries, where environmental standards may be weaker; any risk of pollution gets transferred abroad. Such foreign direct investment might accelerate economic growth in haven countries, while firms that invest can avoid the costs associated with more stringent environmental regulations in their domestic context. In contrast, a pollution "halo" effect means that foreign firms use cleaner energy and thereby improve, rather than worsen, air quality in low- and middle-income settings. A considerable academic literature provides evidence both for and against the haven hypothesis.

One dramatic example, from 40 years ago, illustrates the terrible impact of offshoring production and having another country bear the risks of pollution. This is the explosion at the Union Carbide pesticides plant in Bhopal, India, in 1984, an "acute" event rather than the enduring chronic exposure to well-known pollutants considered later. I have reviewed some aspects of this disaster in previous work (Gatrell & Elliott 2015: 324–6) but revisit it here with more recent evidence.

Figure 7.1 Aerial view of the industrial gas leakage site situated at Bhopal, Madhya Pradesh, India
Source: iStock.

The context in the 1980s was the outsourcing of pesticide production from the United States to India, as set out by Bogard (1989) and Shrivastava's "anatomy" of the disaster (1992). Pesticides were required to support India's Green Revolution, which sought to industrialize food production. The plant's location in India was determined by the need to transfer hazardous production technologies from the Global North to a place where cheap labour was available. Investment in health and safety by the parent company was limited.

The explosion at the plant led to a leak of highly toxic methyl isocyanate and hydrogen cyanide (Figure 7.1). In trying to escape the gas cloud, people ran away, but some 2,000 had died by the following day (Amnesty International claims that up to 10,000 people died within three days: Shetty 2014), and about half a million were exposed to the poisonous gases. Epidemiologists were quick to assemble evidence on the immediate effects on population health, and have followed up cohorts of the affected population over many years. Research by Indian public health doctors six years afterwards found that the rate of miscarriage in the group most exposed to the gases was 24 per cent, compared with 6 per cent in those living further away. Further research determined that respiratory problems lingered ten years later, with the frequency of these declining with distance from the explosion; the most recent epidemiological evidence is reported in De *et al.* (2020). There are valuable stories that continue to be told by those caught up in the disaster. One example conveys much more than do data on the percentages of people who died or suffered subsequently (Box 7.1).

BOX 7.1 A STORY FROM BHOPAL

Aziza, a woman living in Bhopal, tells us:

> I was living with my husband's family at that time. My daughter Ruby was three and my son Mohsin was about eight months old. At about 12.30am I woke to the sound of Ruby coughing badly. The room was not dark, there was a street light nearby. In the half-light I saw that the room was filled with a white cloud. I heard a great noise of people shouting. They were yelling "bhaago, bhaago" (run, run). Mohsin started coughing too and then I started coughing with each breath seeming as if we were breathing in fire. We tried closing all the doors and windows to stop more gas from coming in, but the room was already full of white clouds. My son Mohsin stopped groaning, he fell unconscious. I saw lots and lots of people running, screaming for help, vomiting, falling down, unconscious.
>
> We couldn't talk to each other or even see because our eyes were inflamed. We were wondering what had gone wrong, who had done this. We had no idea

that there was a gas leak from Union Carbide. [On returning home] we saw that the trees had shed all their leaves, which looked as if they had been burnt. At about 8am we heard that people were still running away from Bhopal. My husband arrived home fearing that we had all died.

This extract is one of several told by the Bhopal Medical Appeal (BMA n.d.). Space prevents revealing more but readers can hear the story told by a woman called Saira, which conveys the real horror of the aftermath for one suffering from poverty and abuse as well as the disaster itself.

The disaster was no accident. Bogard (1989: 104) puts it in stark structuralist terms: "Political tradeoffs to enhance legitimacy and global economic demands generated by the capitalist imperative of accumulation narrowed the chances for detecting the dangers at Bhopal." Journalist Apoorva Mandavilli revisited the site in 2018 and reported that, almost 40 years later, the site surrounding the plant remains contaminated with tonnes of waste that leaches into groundwater. Mandavilli says that Union Carbide and its former Indian subsidiary, as well as current owner, DowDuPont, the state and national governments "have all played an endless game of pass the buck" (Mandavilli 2018).

A political economy viewpoint is needed to understand both the immediate horrors of the acute air pollution and the ongoing issues of water contamination from hazardous wastes – 40 years after the explosion.

Monitoring air quality across the globe

In order to assess the impact of air pollution on the health of people in the Global South, and elsewhere, we need good data. A very useful resource of real-time data, for a variety of pollutants (particulates, ozone, sulphur dioxide, nitrogen dioxide), as well as an aggregate index (the World Air Quality Index: https://waqi.info), permits the visual examination of the burden of pollution at dozens of monitoring stations.

Such visualizations tell us nothing about the source of the pollution, which may be local or regional, or transferred across international borders. Various studies have examined the flow of transboundary air pollution. For example, Moon Joon Kim (2019) has studied transboundary air pollutants from China to South Korea, suggesting that these flows from the southwest (particularly around Shanghai) account for just under one-fifth of weekly mean concentrations of large particulates (PM_{10}), although these vary according to season. Some Chinese authors (such as Liu *et al.* 2020) argue that transboundary pollution of smaller particulates (diameter < 2.5 μm: $PM_{2.5}$) from outside China caused 100,000 premature deaths in the country in 2015, though they acknowledge that 90 per cent of all $PM_{2.5}$-related premature deaths result from domestic sources. Meng et al. (2018) examine the movement of black carbon across the globe, movement that is both physical (atmospheric) and virtual (via international trade shipments).

The burden of air pollution is geographically and socially uneven, as revealed in an analysis of India (Chakraborty & Basu 2021). Although data from 2010 show high concentrations of $PM_{2.5}$ in the Indo-Gangetic plain in the north of the country, by 2017 there was a more dispersed pattern, with higher increases in the south. A statistical analysis reveals that there are higher $PM_{2.5}$ concentrations in districts where there are greater proportions of people from Scheduled Castes and poor housing. Further, women and people with disabilities are shown to be increasingly exposed to air pollution, exacerbating the pattern of environmental injustices in India.

Air pollution and health impacts

Small particulates have major impacts on both the cardiovascular and respiratory systems. They can penetrate far into the lungs and bronchial tree. Some writers are in no doubt about the global risks from air pollution. As Nansai and colleagues put it: "Among the many environmental problems affecting human health, the greatest threat is that posed by the inhalation of particles with an aerodynamic diameter of 2.5 μm or less" (Nansai et al. 2021: 2). Such small particles, when inhaled, cause some cancers and respiratory diseases, and lead to over 4 million premature deaths each year, the majority in countries of the Global South. As Figure 7.2 indicates, the (crude) death rates from outdoor air pollution are high in India and Pakistan and parts of north Africa and the Middle East.

It is possible to trace the impact of such small particulate matter on human health by examining international trade and the flows of pollution across national boundaries. Nansai et al. (2021) look at the responsibility of G20 nations for the mortality that results from $PM_{2.5}$, rather than overemphasizing transboundary air pollution. They suggest that about a half of the 4 million premature deaths, mostly in China and India, can be attributed to patterns of consumption in G20 countries. The authors suggest that meetings of the G20 group provide an opportunity to reach an international agreement response to mitigate the global health impacts of small particulates; this links back to the discussion in Chapter 4 on the global governance of health. The G20 group does of course include China and India, as well as other countries (Argentina, Brazil, Indonesia and Turkey) that are regarded as middle-income states.

Sustainable Development Goal 3 (Target 3.9) calls for a reduction in the number of people dying or suffering from air pollution. As noted by India State-Level Disease Burden Initiative Air Pollution Collaborators (2021), there are particular challenges in India, where pollution, particularly from particulates, derives from several sources, namely domestic, industrial and transport-related. Rising energy use, increased urbanization,

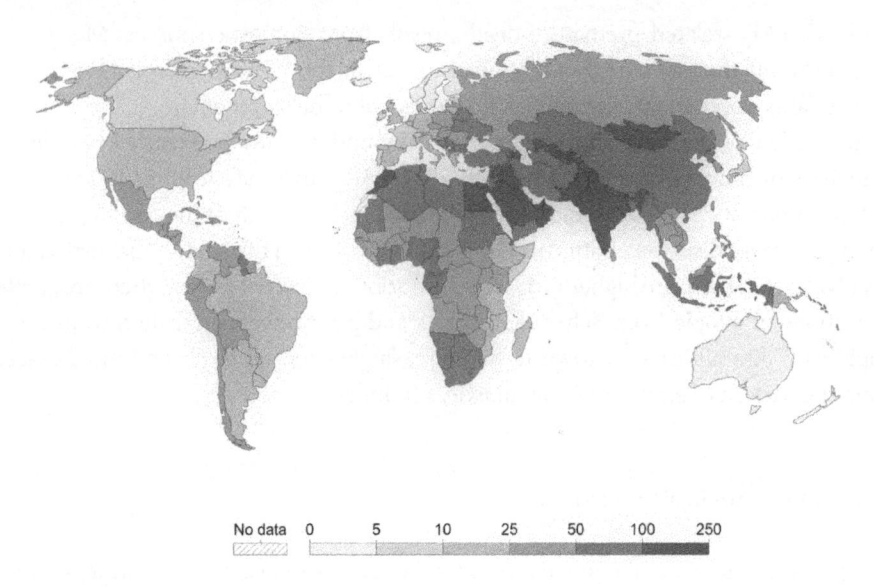

Figure 7.2 Outdoor air pollution death rate, 2017
Note: The number of deaths attributed to outdoor ozone and particulate matter pollution per 100,000.
Source: Ritchie and Roser (2019).

rapid industrialization and growth in the volume of road traffic are all causes. This research group estimates that in 2019 1.67 million deaths (amounting to about 18 per cent of all deaths) were the result of air pollution. The highest concentrations were in northern states. The authors go on to suggest that about $28 billion is lost to the economy because of the premature deaths – losses that also vary from state to state. They conclude that controlling air pollution in India "will not only improve health as envisioned in the SDGs, but will also accelerate the potential to achieve other SDG targets, including alleviating poverty, promoting social justice, enhancing the liveability of India's cities, and reducing the pace of climatic changes" (India State-Level Disease Burden Initiative Air Pollution Collaborators 2021: e34).

A companion paper, also from the Global Burden of Disease study group (Yin *et al.* 2021), looks at geographical variation in air pollution and health among Chinese provinces. As with the Indian study, the authors find very large numbers of deaths (1.24 million) attributable to air pollution in 2017, although the death rate had improved markedly since 1990. By far the worst affected province is Hebei, followed by Hunan, Tianjin and Hubei (see Figure 2.5 for locations).

The health impacts of indoor air pollution, resulting from poor ventilation and the use of coal and other solid fuels (charcoal, crop waste, animal dung) for heating and cooking, have been widely studied, in India and elsewhere. Such pollution and ill health are socially determined, with poor people and women in the Global South worst affected; poor nutrition acts in concert with household air pollution to increase the risks of respiratory and other diseases among vulnerable children. Better-off families can use other fuels, such as liquid petroleum gas. Research in India by Faizan and Thakur (2019) based on a large national sample confirms the associations between type of fuel use and respiratory disease. Exposure to solid fuels (firewood, animal dung) for cooking increases the potential risk of chronic obstructive pulmonary disease (COPD) and asthma among those living in rural areas and those who are socially and economically marginalized (Scheduled Tribes and Scheduled Castes). Crop burning too creates poor-quality air in rural areas.

As this example – and many others – suggests, associations between air pollution and ill health are mediated by social and economic factors, as a political ecology framework clarifies. Véron (2006) spells this out clearly in his study of air pollution in Delhi, India. As he acknowledges, it is the growth in road traffic as well as urban expansion that has generated worse urban air: "Clean air in cities has become increasingly scarce in the course of industrialization and urbanization fostered by capitalist development and by technological change in transportation" (Véron 2006: 2096. Access to clean air is socially patterned, since many poor people are exposed to polluted air when walking, cycling or working near busy roads. Nonetheless, measures to cut air pollution are driven by middle-class interests.

Solutions to the problem of air pollution must be found "upstream". As Gordon *et al.* (2014) suggest, it is a global health problem requiring public health interventions that should involve different actors, including households, local communities, non-government organizations, national and regional governments and international agencies. A similar stance comes from an analysis of the political ecology of seasonal air pollution in northern Thailand (Mostafanezhad & Evrard 2021), where the crop-burning practices of highland farmers are blamed for this, thereby obscuring the structural (historical, political and economic) drivers – as well as the impact of vehicle emissions – of such pollution.

Stephen Graham argues that the political ecology of air is underdeveloped: "The politics and geographies of bad or lethal air in cities remain remarkably peripheral to the huge growth of political-ecological work on the social and technological productions of nature in urban environments that has in other ways been so productive in human geography" (Graham 2015: 193). There is a distinct horizontal and vertical dimension to how clean air is distributed in cities. Graham quotes the Hong Kong anthropologist Tim Choy that "the rich have access to good air while the poor are relegated to the dregs, to the smog and dust under flyovers or on the streets" (203). The elite in Hong Kong colonize skyscraper penthouses to escape polluted air while marginalized groups breathe in the air polluted by heavily trafficked streets.

By way of contrast, some public health authorities think it is up to the individual to take responsibility for minimizing their risks of ill health as a result of air pollution. For example, Carlsten *et al*. (2020) suggest that exposure to air pollution can be reduced by personal choices, by, for example restricting physical exercise outdoors when pollution levels are high, choosing alternative routes to work and taking notice of air quality alerts. A clearer example of victim-blaming might be hard to find.

In an important paper, Saleh *et al*. (2021) draw on the concept of structural violence in illuminating the impacts of indoor air pollution in Malawi. In the village the authors study, solid fuel (wood) is used for cooking. The study used a mixed-methods approach, with personal monitors used to record air pollution – and the results suggest that international standards of exposure to particulate matter were widely breached. Yet valuable qualitative data suggest that, even though the health impacts of chronic smoke exposure were well understood, it was very difficult to reduce exposure to smoke (Saleh *et al*. 2021). As they observe, there is a wider, socio-economic context to air pollution, such context compromising respiratory health.

Monitoring water security and quality

SDG 6 aims to "ensure the availability and sustainable management of water and sanitation for all". The specific target related to drinking water (SDG 6.1) seeks by 2030 to "achieve universal and equitable access to safe and affordable drinking water for all". But the interim report in 2021 on SDG 6 progress (UN Water 2021) makes disappointing reading. One-quarter of the world's population lack access to clean drinking water; almost 30 per cent of the world's population have no domestic handwashing facility; almost half lack safe sanitation; and only 24 countries report that all the rivers, lakes and aquifers that they share with their neighbours are covered by operational agreements.

I considered in Chapter 4 issues of governance and security in relation to global health. Similar issues arise when we look at water resources. Among many definitions of water security, the United Nations suggests it is

> the capacity of a population to safeguard sustainable access to adequate quantities of acceptable quality water for sustaining livelihoods, human well-being, and socio-economic development, for ensuring protection against water-borne pollution and water-related disasters, and for preserving ecosystems in a climate of peace and political stability. (UN Water 2017)

But the UN also refers to transboundary cooperation, whereby states "discuss and coordinate their actions to meet the varied and sometimes competing interests for mutual benefit" (UN Water 2017). However, the UN also notes that achieving SDG 6 is a national responsibility (UN Water 2021). This echoes the debate discussed in Chapter 4 about a Westphalian versus post-Westphalian form of governance, the former prioritizing national interests.

Jepson *et al.* (2017) adopt a relational approach to water security, focusing less on water simply as a material good to be accessed and more on water as a hydro-social process. Here, there is a recursive relation between water and society; for example, modifying water flows and quality affects social life, which in turn affects water use. Their broader approach to water security demands an engagement with politics at a variety of scales, as well as cultural relations. Like others, they link water security to human rights: the rights of individuals and communities to have affordable and clean water to support human capabilities and wellbeing. They offer an example to show that water is not merely H_2O. In Antofagasta, on the Chilean coast, a water supply company diverted mountain water from residential areas in order to serve mining interests. As a result, water has now been supplied by that company from a desalination plant, giving what residents perceive as poorer-quality water. As for culture, water rights must also be linked to what some groups see as a spiritual or sacred duty to protect supplies in order to protect its common heritage for future generations. In essence, "rather than securing water per se, we argue that water security should be about transforming water–society relations to promote human wellbeing and empowerment" (Jepson *et al.* 2017: 50). This relational approach was prefigured by Swyngedouw (2009: 56), for whom "hydro-social research envisions the circulation of water as a combined physical and social process, as a hybridized socio-natural flow that fuses together nature and society in inseparable manners". But Swyngedouw goes further, suggesting that access to water is shaped by the power of money; such access for those living in the Global South is very precarious – and it has health consequences too, as we see later. Having said this, Uitermark and Tieleman (2021), in their study of water infrastructure in Accra, Ghana, argue that it is too simplistic to suggest that unequal access simply reflects colonial history; piped water supplies are not restricted to the elite. Nonetheless, access to connected water supplies remains patchy.

The subtitle of this book is *Geographical Connections*, and few connections are more obvious than those that see rivers cutting across international boundaries. Among many examples, that relating to the Mekong river basin (Figure 7.3) is highlighted in Box 7.2. The river Mekong flows through China, Myanmar, Cambodia, Laos, Thailand and Vietnam, all (with the exception of China) deeply affected by war in the latter half of the twentieth century.

Transboundary effects apply not only across international borders. In China, "polluting thy neighbour" seems to be the case, with provinces concentrating their industrial water-polluting activities in those areas (counties) that are most downstream within the province, thereby shifting the burden of water pollution to neighbouring provinces further downstream (Cai, Chen & Gong 2016) – a kind of geographical spillover effect. Although provincial governments are required to enforce environmental

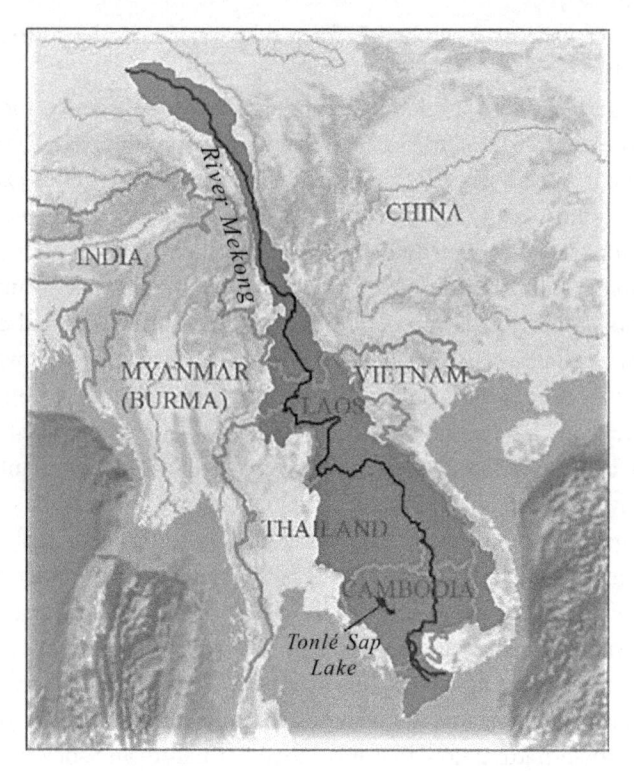

Figure 7.3 The Mekong river basin
Source: www.usgs.gov/media/images/lower-mekong-river-basin-0.

regulations, it appears that their strategy is to devote fewer resources to doing so in downstream regions of the province.

Although water security is of course a pressing issue in parts of the Global North, it is a particular issue for Indigenous communities. Among Native Americans about 6 per cent lack household plumbing, compared with 0.3 per cent of white Americans. But for Indigenous groups water security is both a material and a cultural and spiritual concern, reflecting the same kind of knowledge and environmental relationship that we considered earlier. As US environmental activist Winona LaDuke says: "We are the people who live by the water, pray by these waters, travel by the waters, and eat and drink from these waters. We are related to those who live in the water. To poison the waters is to show disrespect for creation. To honor and protect the waters is our responsibility as people of the land" (cited in Mitchell 2019: 100). As Mitchell (2019) notes, the contrast with Western views that water is, increasingly, a commodity (often bottled) to be bought, sold and controlled could not be more stark.

Nunbogu and Elliott (2021) argue that inequalities in access to water intersect with gender inequalities, stressing that a political ecology of water demands a feminist lens.

BOX 7.2 TRANSBOUNDARY ISSUES IN THE MEKONG RIVER BASIN

All governments in the Mekong river basin are pursuing, or have created, major development projects, particularly the construction of dams for hydropower. Although there is a degree of consultation and cooperation (largely via the Mekong River Commission, established by Laos, Thailand, Cambodia and Vietnam in 1995), "governments in the Mekong view the concerns of neighbouring countries and concerns about negative social and/or environmental impacts, as a nuisance" (Campbell 2016: 261). Further, there is little public consultation or participation in any decision-making.

The major infrastructure projects place huge pressures on the environment and on the population of some 70 million living within the river basin as a whole. There are currently 20 large dams built or under construction in southern China, Laos, Thailand and Cambodia, and, despite the aims of the Mekong River Commission, there is little acknowledgement of the externality effects generated by a dam in one country spilling over to neighbouring countries. Such effects include the prevention of sediment from reaching downstream, thereby impacting soil fertility. The dams affect the migration and supply of fish, with serious consequences for the health and wellbeing of those dependent on fishing. The Mekong River Commission estimated in 2018 that fish stocks would fall by between 40 and 80 per cent by 2040, and sediment by between 67 and 97 per cent (Mekong River Commission 2019).

Sneddon and Fox (2016) take a critical view of transboundary cooperation, which, they say, may worsen river ecology, contribute to resource degradation and disrupt the lives of those who depend on the Mekong. For them, a critical hydropolitics requires a different imagining of the river basin, one that is, in a sense, an actor network involving human actors, organizations and the river itself. As they point out, the establishment of the Mekong River Commission in 1995 failed to focus sufficiently on social ecology (and especially fishing impacts), being more concerned with engineering hydrology. Opportunities for genuine participation by local communities have been limited. The boundaries, they argue, are not just territorial ones but epistemological ones that separate the different knowledges of technical and lay experts. Clearly, there are real, and potential, geographical connections among interest groups living in all regions of the Mekong river basin.

This is because, although women bear most of the responsibility for ensuring water availability (for cooking, drinking, washing) in the home, decisions about water governance in the community tend to be a male preserve. Travelling to find and waiting to collect water fall largely to women and girls (Figure 7.4), and there are accounts of violence towards them either on the way or at water sources. Working with other African scholars (Bisung *et al.* 2015), Susan Elliott used photographs to motivate discussions with eight women

Figure 7.4 Women collecting water near Mzuzu, Malawi
Source: iStock.

living in Usoma, Kenya, on the shore of Lake Victoria. This is a community with no access to safe water and with 42 per cent of the population lacking any toilet or latrine. Pit latrines are regarded as inadequate. As respondent Wintima put it: "You could even see house flies all over, they fly in and out. Even the doors are not there, so there is very little difference between defecating in the bush and this toilet" (Bisung et al. 2015: 211). Local government action is found wanting. Another interviewee, Betty, argued: "If the administration was fair to provide us with water most of our problems will be solved. Sometime we need to talk about the way we have been cheated and neglected over the years" (213). Human agency is constrained by broader political structures, reinforcing inequalities in access to clean water and adequate sanitation.

The same point is made by Workman et al. (2021), who suggest that water insecurity needs to be linked (as a "syndemic": see Box 8.1) with food insecurity, since both are set within the wider environment of structural inequality. Their research highlights a study of nearly 7,000 households in 27 countries in the Global South, which shows that those households with poor access to potable water were also subject to food insecurity. As Figure 7.5 shows, water and food insecurity intersect, leading to malnutrition, stress

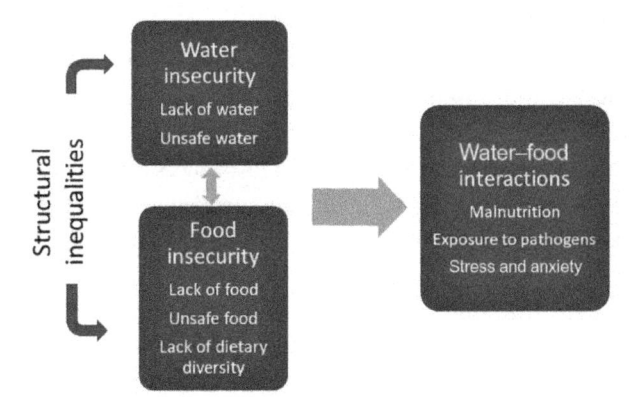

Figure 7.5 Water and food insecurity intersect to cause health problems
Source: Author's own, based on Workman *et al.* (2021: fig. 1).

and anxiety, and exposure to infection. The next section discusses these health effects in more detail.

Water quality and health effects

Detailed estimates by a large collaborative group have been made both of access to drinking water and sanitation facilities and the likely impact on childhood diarrhoeal disease (Local Burden of Disease WaSH Collaborators 2020). Such estimates are made for 88 low- and middle-income countries, at subnational scales. Access to drinking water is classed as piped, protected (bottled water, tanker truck) or unprotected (surface water), while access to sanitation is via sewer/septic tank, latrine or no facility. Among all LMICs access to piped drinking water had improved between 2000 and 2017, from 40 per cent to 50 per cent, but this varied geographically, and there were large numbers of households with poor access to piped water or safer sanitation facilities. For examples, over a quarter of a million people in Harare, Zimbabwe, lacked access to sewers or septic tanks. Populations relying on open defecation were concentrated in parts of west and central Africa. But in parts of southeast Asia, notably Cambodia, there was evidence of improved sanitation, and a reduction in unequal access within that country. The authors note that, "alongside economic growth nationally, the autonomous Phnom Penh Water Supply Authority has been credited with substantial expansion of Cambodia's urban piped water supply" (Local Burden of Disease WaSH Collaborators 2020: e1172).

What are the consequences for childhood diarrhoea? The authors estimate that, in 2017, 182,000 deaths of children aged under five could be attributed to unsafe sanitation. A companion paper has explored and mapped access to oral rehydration therapy to treat diarrhoea (Local Burden of Disease Diarrhoea Collaborators 2020), indicating that over

6.5 million children went untreated in 2017. There are substantial between-country and within-country inequalities. Results suggest that the availability of ORT had increased between 2000 and 2017 in Cambodia, Vietnam, India and Rwanda, but had declined in Sudan and South Sudan. Given that the packets of glucose and electrolytes require dissolving in clean water, and that evidence suggests access to ORT is higher among wealthier households, these inequalities are hardly surprising. As with the WaSH collaboration (Local Burden of Disease WaSH Collaborators 2020), this paper gives considerable spatial detail on the geographical variation within particular subregions.

Somewhat wistfully, the paper concludes that "we are surprised to see low use [of ORT] after so many years of programmes in many countries, especially those with high diarrhoeal burden" (Local Burden of Disease Diarrhoea Collaborators 2020: e1048). As I noted in Chapter 2, improving primary healthcare delivery and, crucially, addressing the upstream social and economic determinants of health need to go alongside improvements to water and sanitation and the use of ORT. These offer the best path to reductions in child mortality.

Haiti proves a salutary example of links between water quality, sanitation and disease, notably the bacterial disease cholera. By way of background, in 2004 a UN peacekeeping force was established in Haiti to supervise elections and maintain order. The UN mission proved deeply controversial, not least because of many allegations of violence and sexual abuse committed against the local population. But following an earthquake in 2010 its presence led to an epidemic of cholera.

The earliest cases of cholera in 2010 were traced to the river Méyé, a tributary of the country's longest river, the Artibonite. A drainage pipe was sending raw waste from a bathroom on the UN site directly into the river, which was used as a domestic source of water. It subsequently transpired that the particular strain of cholera was from Asia, and, although it was disputed by the UN, it became evident that UN personnel from Nepal were responsible for this pollution, and at the time one in ten Haitians with cholera died. Subsequently, between 2010 and 2019, there were about 800,000 cases of the disease. Since then the numbers have dropped, and in 2020 the country reported no cases, because of intense activity to establish cholera clinics and provide for the rapid dissemination of ORT when cases came to light.

Paul Namphy, a cholera coordinator for Haiti's Water and Sanitation Authority, cautioned in 2020 that acute diarrhoea and typhoid were more of a risk than cholera (Kushner 2020). As seen elsewhere in this section, structural problems of water supply and sanitation – upstream factors – are needed to limit the spread of waterborne disease in the country. Before the 2010 earthquake only two-thirds of Haiti's population had access to water from a pipe or well, and only 17 per cent had access to adequate sewerage or a latrine. These figures have improved somewhat, although in 2020 more than half the country's rural population still lacked access to drinkable water and only about one-third had access to basic sanitation (World Bank 2020). Griffiths *et al.* (2021) adopted a spatial analytic approach in their investigation of over 5,000 cases of cholera in one subregion of Haiti between 2015 and 2016. Their statistical analysis suggests that those localities lying close to rivers or close to unimproved water sources had the highest incidence of cholera. There may be some value in identifying "hotspots", but whether we require such quantitative analysis to demonstrate the links to contaminated water is a moot point. In 2022 Haiti remained a country in severe crisis, not only because of further outbreaks of cholera but as a result of corruption and gang violence.

Other waterborne diseases, notably "neglected" tropical diseases (NTDs), are associated with poverty (Hotez 2008). SDG 3, "ensuring healthy lives and promoting well-being for all at all ages", includes an ambitious, doubtless unrealistic, target to end NTDs by 2030. Two such NTDs, onchocerciasis and schistosomiasis, are discussed here.

Onchocerciasis, commonly known as river blindness, is caused by a parasitic worm carried by blackflies, the disease vector. When the flies – which breed near rivers – bite they transmit the larvae of the worms (genus *Onchocera*), which lie under the skin and cause body disfigurement, and blindness when they migrate to the eye. According to Hotez, it has a global prevalence of about 37 million, with the highest prevalence in sub-Saharan Africa, where the rate of blindness in some villages can be as high as 10 per cent. Since river blindness is most common in adulthood, the hardest hit are householders who can no longer contribute economically, so fields may be left untended and food shortages result. Some success in controlling the disease is attributable to the use of the drug ivermectin, promoted by the African Programme for Onchocerciasis between 1995 and 2005. However, despite this, some countries continue to show high prevalence; in parts of Cameroon, for example, it is over 44 per cent (Nji *et al.* 2021). Partly this is a consequence of limited community participation in the organization and planning of health delivery, but other contextual factors are involved. In the farming communities in the Meme river basin, Cameroon, studied by Nji *et al.* (2021), workers spend many hours in cocoa plantations and are not at home when medicines are distributed. Boreholes and wells dry up during the summer, so people go to the rivers to fetch water or to wash, exposing themselves to the blackflies.

This qualitative research in Cameroon is mirrored in Liberia, where narratives collected by Dean *et al.* (2019) reveal how the illness and disability experienced by those with onchocerciasis (as well as others NTDs) creates upheaval, or "biographical disruption". Some women report being fearful of moving too far from their homes because of worries about safety. For example, Janet's mother joined her to a rope to guide her from the house to follow so she could defecate nearby, as she was worried about being attacked if she moved too far. She says: "I am used to it, I am used to blindness. [I could] be walking around – I even know the road, but it scares me, it is dangerous to me because I am not seeing what is ahead of me so that['s] what cut it [her movement] short" (Dean *et al.* 2019: 14). As with many other health issues in countries of the Global South, gender intersects with other asymmetries of power to structure ill health, and to address such issues psychosocial support needs to go hand in hand with medical interventions. For the authors, "the narratives presented here are deeply grounded within the unique political and historical trajectory of Liberia (a nexus of conflict, colonialism and aid-dependency)" (Dean *et al.* 2019: 16).

Like onchocerciasis, schistosomiasis (also known as bilharzia or "snail fever") is most prevalent in sub-Saharan Africa; globally, the prevalence is about 207 million (Hotez 2008). The disease is caused by parasitic worms that live in freshwater snails; the snails pass the larvae, which can penetrate the skin, thereby transmitting the parasite. The adult worms lay eggs that are shed through urine or faeces. The disease causes bleeding and anaemia and can lead to bladder cancer. The drug praziquantel (for deworming) has helped control disease prevalence. Spatial statistical analysis enables detailed mapping of the prevalence of schistosomiasis, as, for example, across east Africa (Schur *et al.* 2013). Whether such maps are actually used to plan and control interventions – as the writers hope – remains to be seen.

In the absence of any vaccine, the US Centers for Disease Control and Prevention (CDC) recommend drinking safe water and avoiding swimming in freshwater, advice that surely does little to help those for whom "safe" water is impossible to come by. Inevitably, there are strong links between infection risk and poverty. Houweling *et al.* (2016) report a study of Ugandan teenagers that suggested the odds of infection were 55 times greater in the poorest, compared with the wealthiest, households. In southwest Nigeria the prevalence of one form of schistosomiasis was 70 per cent for households in which income was less than $600 but only 1.5 per cent if income was greater than $1,600 (Savioli *et al.* 2017). Although treatments are available, the expansion of control programmes in Africa is progressing slowly, with disease transmission continuing in many rural areas, hindering economic development and human wellbeing. Clearly, there is a reciprocal relationship: poverty is an upstream cause of the disease, while the disease, through limiting work, shapes poverty. Such a relationship is common to all NTDs. Well-intentioned partnerships, such as the Global Schistosomiasis Alliance (Savioli *et al.* 2017), established to advocate and push for greater dissemination of prevention and treatment, may have limited value unless the structural socio-economic determinants are fully addressed.

A different type of water contamination is that attributable to arsenic. In Bangladesh, with a population of 165 million, it is estimated that up to 50 million may be drinking arsenic-contaminated groundwater taken from tubewells (Figure 7.6). Such wells were seen as a safe alternative to surface water, which was possibly contaminated by cholera and other bacteria. The arsenic derives from rocks in the Himalayas, and extensive river systems carry sediments, which are deposited in the floodplains with the arsenic leaching into the groundwater. The issue first came to light in 1993, but by 2005 it was clear that arsenic contamination was present in the drinking water in all but two of the country's 64 districts. Chronic exposure causes arsenicosis, leading to skin disease and cancer. Contaminated water used for crops such as rice means that arsenic can find its way into

Figure 7.6 A tubewell in Bangladesh where water is contaminated with arsenic
Source: iStock.

the food chain. Ahmad, Khan and Haque (2018) provide an excellent background to the problem.

Several studies have examined, and mapped, spatial variation in arsenic contamination and sought to link this to health outcomes, such as adverse pregnancy outcomes (Sohel *et al*. 2010) and skin cancer (Choudhury *et al*. 2018). Farhana Sultana (2006, 2012) has studied this in the context of Bangladesh, using a political ecology framework. We noted earlier the gendered nature of access to water in LMICs, and, as she demonstrates, this is very much the case in Bangladesh. The experiences of suffering are, she argues, gendered, with women expected to continue household chores even though living with arsenicosis; but this intersects with poverty, since wealthier women can afford help. Further, since the disease can manifest itself with skin lesions, a girl's "marriageability" may be limited because of the stigma. The intersection with poverty manifests in other ways; wealthier households can dig deeper wells, below the contaminated layers. As Sultana (2006: 374) puts it: "Social differences interact with the spatial variability of arsenic contamination to produce spaces of power of some over others who require access to water."

Sultana also points out the political dimension, noting that the country's capital, Dhaka, is not in an area with water contaminated by arsenic and that, had it been so, then action to address the problem might have been speedier. Criticism of government (in)action continues. A report by Human Rights Watch (HRW) in 2016 concludes: "The government acts as though the problem has been mostly solved, but unless the government and Bangladesh's international donors do more, millions of Bangladeshis will die from preventable arsenic-related diseases" (HRW 2016).

Arsenic contamination affects mental as well as physical health (Sultana 2011, 2012), with the stigma carried by those known to have arsenicosis contributing to their emotional distress, especially if they are perceived (wrongly) by others to be contagious; such distress is borne especially hard by women. Anticipating Sultana's work, this was explored in detail by Hassan, Atkins and Dunn (2005), who report distressing stories from those whose skin is discoloured as a result of the poisoning. For example, Taslima (ten years old) says: "Nobody sits beside me in school. They do not like to talk with me, and do not share books. Nobody likes to play with me in school. When I play, some children shout 'Don't touch her, don't play with her, she's got arsenic'. I will not go to school" (Hassan, Atkins & Dunn 2005: 2206).

Such links with mental health are seen in other settings. Many studies in low- and middle-income countries have shown how water *in*security – poor access to clean, accessible, affordable and safe water – affects mental health. Research using a large sample of women in eastern Ethiopia (Brewis *et al.* 2021) reveals that depression and anxiety are shaped by water insecurity but that this is mediated by household income. However, crucially, it is women's sense of unfairness and injustice in access to water resources that has a significant mediating effect on wellbeing. Access here means who owns wells and pumps, how far the household is from shared water sources and what transport is available for carrying the water.

Concluding remarks

The focus in this chapter has been on the (lack of) clean air, clean drinking water and adequate sanitation in large parts of the Global South, and on the health consequences. Of course, this is not to downplay the serious impacts on human health of polluted air in the Global North (whether in heavily trafficked city streets, or because of the recent forest fires in parts of Australia and in California). Nor too is it to minimize the lack of access to potable drinking water in parts of the Global North; recent research by Meehan *et al.* (2020) notes that almost half a million households in the United States lack piped water, while the crisis in Flint, Michigan, that resulted from lead contamination in water supplies has been well documented and subject to geographic analysis (Hanna-Attisha *et al.* 2016).

Like air, water is mobile, but it is far from ubiquitous and is a commodity to be bought and sold. The security of both, water in particular, has been explored above, and issues of governance and security take us back to those issues covered in Chapter 4. Crucially, as we will see in Chapter 9, the mobility of water, in the form of flooding, is closely linked to climate change.

The concept of a syndemic, introduced in the following chapter, is important in understanding associations between air and water quality and human health, since both interact with poverty and food security in shaping population health. As elsewhere in this book, while accepting that classic epidemiological studies have a part to play, more valuable are the stories told by people living with polluted air or insecure water supplies. Such stories, set within the framework of political ecology, bring home very clearly the structural violence endured by many people living in the Global South.

Further reading

With Susan Elliott I have reviewed some of the relevant literature on both air and water pollution and how their impacts play out in different places, including parts of the Global South (Gatrell & Elliott 2015: ch. 11). On air pollution and health, a good general reference is Kessel (2011). For studies of particular world regions or specific countries, I recommend a literature search using PubMed (https://pubmed.ncbi.nlm.nih.gov).

On water politics, see the chapters in Sultana and Loftus (2019). The geopolitics of water security in the Mekong river basin (and many other topics) are discussed in Doyle and Risely (2008).

Sultana's research (2011, 2012) on arsenic contamination is essential reading. On water, sanitation and hygiene ("WaSH"), see Bisung *et al.* (2015), and on the crucial gender dimension, see Abu, Bisung and Elliott (2019).

8
INFECTIONS ON THE MOVE

> There is an emerging political ecology of disease in the twenty-first century in which the relationship between globalization, urbanization, and pathogenic organisms is being recast, with potentially devastating consequences.
>
> Gandy (2008: 182)

As clarified in the preface to this book, its subtitle, *Geographical Connections*, can be interpreted both in terms of the connections or flows that occur in geographic space and the relations between the environmental and social factors that shape disease incidence and unequal health. Two earlier chapters have dealt separately with the mobilities of populations and material goods, and I turn now to the spread – but also the incidence – of infectious disease. Of course, the movements of materials, people and pathogens are all interrelated.

The emergence of a new coronavirus in late 2019/early 2020 has, as Cambridge geographer Matthew Gandy predicted in 2008, brought into the sharpest focus the impact of a new disease (Covid-19), and I consider here, briefly, the origins and wider determinants of the infection. I devote most of my attention to the differential impacts on (and within) nation states, particularly in the Global South, as well as inequities in how the disease has been managed through vaccination. Discussions of global vaccine availability and equity take us back to issues of global governance discussed in Chapter 4.

A key theme in the spread of the virus (SARS-CoV-2) that has transmitted Covid-19 relates to the networks of transmission through international movement, and I consider how changing patterns of air travel have enabled the spatial spread of infections – including newer strains of influenza. But I also consider other emerging and re-emerging infections, including an earlier coronavirus (SARS), as well as Ebola, for which mobilities across borders were less related to air travel but, rather, to road and track. The late Paul Farmer argued, in his last book, that the failure to deal with Ebola can in part be traced back to colonial exploitation and injustice. As he puts it, in west Africa "disease containment was a priority but care was not" (Farmer 2020). Much the same was true about Covid-19 in many parts of the world.

Earlier he was also careful (Farmer *et al.* 2013) to make the point that empirical research on disease requires not merely comparisons between countries but detailed data collection on local and regional variations in incidence. It is therefore appropriate to begin by describing briefly some of the geographic tools and data that help monitor disease incidence and spread.

In order to describe and analyse geographies of infectious disease, we require spatially referenced data on populations at risk, disaggregated at the finest possible scale. These data sets are necessary, if not sufficient, for disease surveillance. They are usually aggregated to administrative districts. However, population censuses are often infrequent and error-strewn, and it is now possible to use remotely sensed imagery, including data on land cover, to produce detailed and up-to-date estimates of population (see, for example, Bharti *et al.* 2016). There are many organizations and individuals engaged in the use of geographical information systems/science for monitoring the spatial distribution of diseases and their determinants. Among these, the WorldPop team based at the University of Southampton is particularly notable. The group has produced a variety of population data for 100 m grid squares, and the data sets also include (for example) estimates of pregnancies and live births in South America (as part of research to examine the prevalence and impact of Zika virus: see below, pp. 156–7). In most cases the data sets constructed by WorldPop (currently almost 45,000) are available to download (see www.worldpop.org).

Geographic data collection and analysis for infections "on the move"

It is a truism that diseases spread from place to place because of human mobility, and this has always been the case, whether we consider the Black Death, the 1918/19 influenza pandemic or the spread of HIV in the last quarter of the twentieth century (see Gatrell 2011: 122–3 for an overview). It is also true that the volume and speed of transport has led to time–space convergence (Figure 1.4), at different spatial scales but most notably intercontinentally.

As a result, we need data and methods that allow us to quantify such spread. In some cases, the data are so rich, even in historical contexts, that we can track the movements of infectives over time and space (see Cliff *et al.* 2009 for outstanding examples). This is of course almost impossible when dealing with common infections, such as malaria (see below, pp. 156–60), so researchers have used a variety of other data to represent, or model, flows that are likely to reflect disease spread. Such data include that on migration drawn from population censuses. For example, Sorrichetta *et al.* (2016) have estimated internal human migration flows between administrative subdivisions for all malaria-endemic countries in Africa, Asia and Latin America. As a companion paper has it: "Mapping the routes of parasite dispersal by human carriers will allow for additional targeted control by identifying both the regions where imported infections originate and where they may contribute substantially to transmission" (Wesolowski *et al.* 2012: 267). As a result, if we know about malaria prevalence in one region, we can predict the likely transmission to different areas. Research by Strano *et al.* (2018) has explored the connectivity of the road network across the entire African continent, linking this with both demographic data and malaria prevalence to identify those regions most closely linked to areas of high disease burden. Their maps highlight how such linkages cut across national borders, suggesting that efforts to control malaria have to be addressed by neighbouring countries.

In other research involving the Southampton WorldPop group, Tatem, Rogers and Hay (2006) have linked monthly climate data for many of the world's major airports with data on airline traffic and seasonal malaria prevalence to estimate which of the air routes were most likely to lead to mosquito importation. For example, flights from Abidjan in Côte d'Ivoire to Paris, and Accra in Ghana to Amsterdam, were suggested as high-risk routes for the import of mosquitoes carrying malaria parasites. Similar research has been undertaken by Lai *et al.* (2018), on the seasonal risk of dengue fever being imported by air to 165 Chinese cities from nine countries in southeast Asia. Results suggest that, although large cities located on the coast (Beijing, Shanghai and Guangzhou) are at high risk of disease importation, inland cities (Chengdu and Wuhan) are emerging as potential locations for dengue outbreaks.

Another source of data is that from mobile phone records. Given the increasing and near-universal penetration of mobile phones across the globe and the means of estimating user locations in real time, such "big data" offer the possibility of estimating population counts at a detailed spatial scale. Wesolowski *et al.* (2012) use such data to model the changing locations (travel behaviour) of 15 million people in Kenya in 2008/09. Detailed small area data were available on malaria prevalence, and so estimates can be made of the likelihood of cases being imported to specific places.

Other researchers have acknowledged that, if we want to assess people's exposure to infections, we need to understand their mobility (see Chapter 5 for an extended discussion). An excellent example is work undertaken by Fornace *et al.* (2019), who, instead of using mobile phone data, used GPS tracking to monitor the local movement of 243 people in forested areas of north Borneo. Linking these data to that on mosquito ecology and land use, they are able to assess the likely risks of biting and infection with a malaria parasite. Since much malaria control relies on use of bed nets and indoor spraying, the study is important in demonstrating that outdoor exposure is also highly significant.

Geographies of neglected tropical diseases

The devastating health burden on people in the Global South caused by HIV, tuberculosis and malaria is well known. Indeed, SDG 3.3 refers explicitly to these three. But it also refers to neglected tropical diseases – neglected, that is, relative to the attention and funding devoted to the others. Farmer *et al.* (2013: 319) are rightly sceptical about

Table 8.1 The major neglected tropical diseases

Disease	Global prevalence	Population at risk
Ascariasis (roundworm)	807 million	4.2 billion
Trichuriasis (whipworm)	604 million	3.2 billion
Hookworm	576 million	3.2 billion
Schistosomiasis (snail fever)	207 million	779 million
Lymphatic filariasis (elephantiasis)	120 million	1.3 billion
Trachoma	84 million	590 million
Onchocerciasis (river blindness)	37 million	90 million
Leishmaniasis	12 million	350 million

Source: Hotez (2008: 5).

the word "neglected": "The essential truth is that all diseases of the poor are perforce neglected." The previous chapter considered two of these: schistosomiasis and onchocerciasis. The global prevalence, and populations at risk, of many NTDs is quite extraordinary, as revealed by data published in 2007 (Table 8.1).

Lymphatic filariasis is transmitted by mosquitoes, onchocerciasis by blackflies, and leishmaniasis by sandflies. An understanding of the geographical epidemiology of all of these requires the adoption of an approach that draws on disease ecology (see above, pp. 2–6) and, more generally, political ecology. They demand a syndemic perspective (Box 8.1). All are affected by changes in temperature, rainfall and humidity, and these impacts are considered more fully in the following chapter. Deforestation, desertification and soil erosion all impact vector-borne disease, as we see in the next section on mosquito-borne diseases, including malaria and dengue. In addition, it has long been established that large infrastructure projects alter ecologies (Tallman *et al.* 2022). For example, projects to control water supplies, such as dam construction, change the ecology of snail populations and hence contribute to elevated risks of schistosomiasis; this is because dams reduce the numbers of migratory river prawns that eat the snails.

We should note that attempts to deal with NTDs have been seriously disrupted by the Covid-19 pandemic (below, pp. 166–72), which will clearly have an impact on the extent to which the SDG target that addresses NTDs can be met.

BOX 8.1 A SYNDEMIC PERSPECTIVE

The medical anthropologist Merrill Singer introduced what he called a "syndemic" perspective on disease, which bears a close relationship to a political ecology framework (above, pp. 6–10). According to Singer (2010: 25–7), this perspective means that diseases do not occur in isolation; rather, they interact or connect synergistically or relationally with other health conditions (what are called "co-morbidities" in medical literature). Although some current disease surveillance tends to focus on one disease, such a silo approach "does not reflect disease burden at a community level, where multiple infections cluster together in the same community and in the same individuals, mostly the very poor" (Bardosh *et al.* 2017: 10).

In addition, social and environmental conditions interact to set the stage for disease to appear, so we need to think of a biosocial environment as the backcloth for disease prevalence. Global changes in climate, land use and agriculture, large-scale infrastructure projects, urbanization, population movement, poverty, military conflict and drug resistance are all linked in a complex dynamic system. But Singer also wants to acknowledge that environments, built and natural, reflect social inequalities. The unequal distribution of power, coupled with discrimination, stigmatization and other forms of structural violence, all determine disease distribution. Although this is as true for some in the Global North (for example, associations between HIV, tuberculosis and poverty in late twentieth-century New York City), it is a feature for many in the Global South.

Singer subsequently broadened the concept to include the role played by climate change; he therefore speaks of "ecosyndemics" and the "coming plagues" of the twenty-first century. As we see later, one such "plague" arrived in early 2020.

Although a syndemic approach is attractive, it has much in common with a structuralist approach to health, and its relational perspective calls to mind the assemblages outlined earlier (pp. 8–10).

Singer (2010) is well worth reading. See also the journal *Infectious Diseases of Poverty* for other examples of close relationships between one disease and others, as well as a special issue of *The Lancet* (www.thelancet.com/series/syndemics). Anyone interested in an historical perspective on syndemic infectious disease (links between influenza, tuberculosis and social environments) among Indigenous peoples in Canada should consult Herring and Sattenspiel (2007).

In an important paper on neglected tropical diseases, Theobald *et al.* (2017) make the point that gender has been neglected. They note that gender affects vulnerability to, experience of and access to treatment of NTDs. Further, it intersects with other dimensions of inequality, notably poverty, but also ethnicity, disability and age. These features are elaborated in a wide-ranging review of another NTD: leishmaniasis (Grifferty *et al.* 2021). The skin disfigurement (especially facial scarring) that results from this disease means that women are stigmatized and discriminated against, and suffer poor mental health and reduced employment opportunities and hence income. As suggested in Box 8.1, leishmaniasis intersects with other conditions, notably HIV, complicating diagnosis and healthcare.

Others have linked the prevalence of NTDs in some countries to a history of colonialism. Nigeria provides a compelling example, the country accounting for about a quarter of all NTDs in Africa (Hotez 2008 has referred to the country as "ground zero" for NTDs). Given high rates of HIV and sexually transmitted diseases, particularly in rural areas and among women, the country bears all the hallmarks of a syndemic pattern. Better healthcare in urban areas is in part a legacy of British colonialism, which sought both to protect the health of settlers there and marginalize the role of women, who had previously enjoyed some degree of autonomy (Maju *et al.* 2019). In rural areas women are the main contributors to agriculture, giving them more contact with soil and water, the sources of NTD vectors.

Political and military conflict, and the inevitable population displacement that results, also have an impact on the spread of infectious disease. Cliff *et al.* (2009) have given an historical perspective on this (since the Second World War), with a range of examples, including the Korean and Vietnam conflicts (1950s and 1960s respectively). In the twenty-first century Sharara and Kanj (2014) have documented the impact of the Syrian civil war on the epidemiology of measles, poliomyelitis and leishmaniasis. Having previously been eradicated, new cases of polio have emerged, the virus lodging in raw sewage that has been pumped directly into the river Euphrates, with no decontamination with chlorine. Rates of immunization against polio and measles have fallen dramatically because of the destruction of health infrastructure and the collapse of the healthcare system, as more recent evidence suggests (Alhaffar & Janos 2021). Neighbouring Lebanon has suffered similar infectious disease outbreaks as refugees have moved across the border. Berry and Berrang-Ford (2016) have explored the association between the incidence of leishmaniasis across 54 countries and data on armed conflict in the same countries. Those countries reporting high levels of conflict were significantly more likely to report high incidence of leishmaniasis, notably in Sudan, South Sudan and Ethiopia. As the authors note, the association is not necessarily attributable to the direct effects of armed conflict; rather, the wider socio-political situations determine a country's ability to prevent and manage such disease during such conflict.

Political ecologies of mosquito-borne diseases

It is well known that the direct cause of malaria is a bite from a mosquito that has been infected by parasites of the genus *Plasmodium*. Less well known, perhaps, is that the direct causes of dengue, Zika (Box 8.2) and chikungunya are also bites from mosquitoes (of the genus *Aedes*). Among those three, dengue is by far the most common, the WHO estimating that almost 4 billion people are at risk (in South America, sub-Saharan Africa, and south and southeast Asia), with about 390 million infections reported each year. The *World Malaria Report 2022* suggests there were almost 250 million cases of malaria and well over 600,000 deaths in 2021 (WHO 2022b).

As we have seen elsewhere in this book, there are wider determinants of these infections – determinants that are both environmental and social. We need to draw on disease ecology to understand why, and where, people get infected, and on political ecology to understand the deeper structures of spread and infection. We need to look also at proximate factors, such as the stagnant pools of water where female mosquitoes lay their eggs (Figure 8.1).

BOX 8.2 ZIKA VIRUS

Zika virus is transmitted primarily by bites from infected *Aedes* spp. mosquitoes; consequently, as with dengue, the same issues of population risk from pools of stagnant water and poor sanitation arise. The virus first came to the world's attention in 2015, when, in northeast Brazil, it appeared that there was a connection between Zika infection and the birth defect known as microcephaly (a smaller than normal head).

Research undertaken by Messina *et al.* (2016) has identified areas where the environmental conditions (temperature, precipitation, humidity) are supportive of Zika emergence, with Brazil, Colombia and Venezuela being identified as areas of high risk. Over 5.4 million births occurred in 2015 within these and neighbouring countries, but, as the authors acknowledge, not all people living in the risk areas are likely to be exposed to Zika virus.

The risk of microcephaly is spatially and socially uneven. In Recife, northeast Brazil, the 1 per cent of births that were microcephalic in 2015/16 occurred in the poorest districts, where garbage collection was sporadic and sewage systems were inadequate. "The microcephaly outbreak in Recife disproportionately affected the newborns of women who were of black or mixed race, had low levels of education, were dependent on public health care services, and lived in the poorest areas" (Souza *et al.* 2018: 469). Women were advised against becoming pregnant, but "framing Zika as a problem of individual behaviour change – of cleaning backyards or wearing condoms – ignores the role of the state as a caregiver, of civil society as the glue of social change" (Bardosh 2020: 23).

As with any virus, mutations can trigger further outbreaks of Zika infection, as research reported in 2022 suggests (BBC News 2022a).

For a thorough consideration of Zika in relation to global health, the chapters in Bardosh (2020) are an important resource. See also Wenham (2021) for an account of Zika as a matter of global health security for women's lives.

Harris and Carter (2019) look at all three of the diseases transmitted by *Aedes* mosquitoes, in a study set in rural Ecuador. The risk of infection is unequally distributed, with people who have poor access to adequate and safely stored water supplies being most vulnerable. Interviews confirmed that people understood these risks, but the authors' fieldwork observed many uncovered plastic containers for collecting rainwater. As one woman who was interviewed acknowledged: "Yes, sometimes they [uncovered receptacles] can be hazardous, because of the insects that can grow there, but what can you do if they don't give us water every day?" (Harris & Carter 2019: 333). The water

Figure 8.1 Stagnant water as a breeding ground for *Aedes* mosquitoes
Source: iStock.

supplied by local authorities is considered both unreliable and dirty, and so rainwater is preferred for washing. The authors are careful not to overstate the importance of structural factors in determining risk; nonetheless, factors such as an inadequate water supply constrain human agency – local knowledge and understanding of risks – and therefore exposure to disease.

Bayona-Valderrama, Acevedo-Guerrero and Artur (2020) have expanded on this in an urban context, examining both high-income and low-income neighbourhoods in Maputo, Mozambique. There, intermittent supplies mean that water has to be stored; households in higher-income areas use closed containers while those in poorer areas use uncovered buckets and barrels. The *Aedes* mosquitoes are attracted in particular to stagnant water that contains organic matter and is of an optimal temperature (about 30°C) for the mosquitoes to develop. Again, it is a political ecology approach that informs this research: "Power relations influence the flow and pressure of water throughout different neighbourhoods, and the ways in which communities outside of formal networks manage to obtain water" (Bayona-Valderrama, Acevedo-Guerrero & Artur 2020: 187). The interactions of humans and non-humans (mosquitoes, stagnant water, etc.) call to mind the assemblages that were introduced earlier.

As suggested in this study, dengue outbreaks are particularly likely to occur in densely populated, poor urban areas, where water that collects in buckets, pots, discarded containers, open drains and used tyres provides perfect breeding grounds. Mulligan,

Elliott and Schuster Wallace (2012) note that, with increasing rates of urbanization, it is likely that the infection will spread beyond areas of deprivation. Their fieldwork in Putrajaya, the planned capital city of Malaysia, confirms this. Imagined as an "intelligent garden city", the risks of dengue infection were ignored. As one public health official put it: "Malaysians like pots with lots of water in the gardens ... and you get a lot of that, thinking it's very beautiful. You also get people collecting rain water, and more often than not, the dengue mosquito" (Mulligan, Elliott & Schuster-Wallace 2012: 617).

Political ecological approaches to dengue can go hand in hand with detailed geographical data collection and analysis. A good example is research in Rio de Janeiro, Brazil, by Carvalho, de Avalar Figueiredo Mafra Magalhães and de Andrade Medronho (2017) that mapped over 30,000 confirmed cases of dengue in 2011/12. Hotspots occur in and near favelas, suggesting an association between incidence and social vulnerability. The authors speak of the structural violence that is made visible through such hotspots. Looking at the country as a whole, Lowe *et al.* (2020) argue that rapid development in Amazonia has created settlements lacking access to piped water and sanitation. Land use changes, along with unmanaged urbanization and increases in mobility, are leading disease vectors to increase their spatial range. Research suggests that road construction that connects urban and rural areas enables the spread of infection. This is demonstrated vividly by Lana *et al.* (2017), who show how the increasing connectivity of the western state of Acre in Amazonian Brazil to other parts of the country has led to dengue infection in the state. Increases in airline traffic to Acre, along with road developments in the state that have improved the interconnectivity of the state's municipalities, have driven infections.

The spread of dengue across Brazil, into Amazonia, is echoed for malaria. Industrial activity and agroforestry have led to increases in migration from other parts of Brazil. Such mobility helps the disease diffuse across geographical space because people who move from an endemic to a non-endemic region carry the parasite and transmit the disease when an *Anopheles* mosquito bites them and then feeds on other humans. "Historically, the Brazilian malaria landscape was largely shaped by what has been referred to as 'frontier malaria' where new infections were driven by non-immune populations settling for economic and political purposes" (Lana *et al.* 2021: 2). More recently, malaria incidence has declined in many parts of Brazil, but it shows considerable spatial heterogeneity; 80 per cent of notified cases in 2018 were found in just 25 (out of 5,570) municipalities, almost exclusively in the Amazon states, representing 3.1 per cent of the population. Case rates attributable to two species of the *Plasmodium* parasite are mapped in Figure 8.2.

Malaria is no respecter of international borders. Political and economic instability in Venezuela, for example, along with environmental degradation resulting from illegal mining in Bolivar state (neighbouring Brazil), has driven population movement, leading to an increase in malaria cases in parts of Brazil and other countries in South America. The same features characterize other parts of the world. For example, outbreaks of malaria in Laos, in southeast Asia, are attributed to deforestation, development projects and

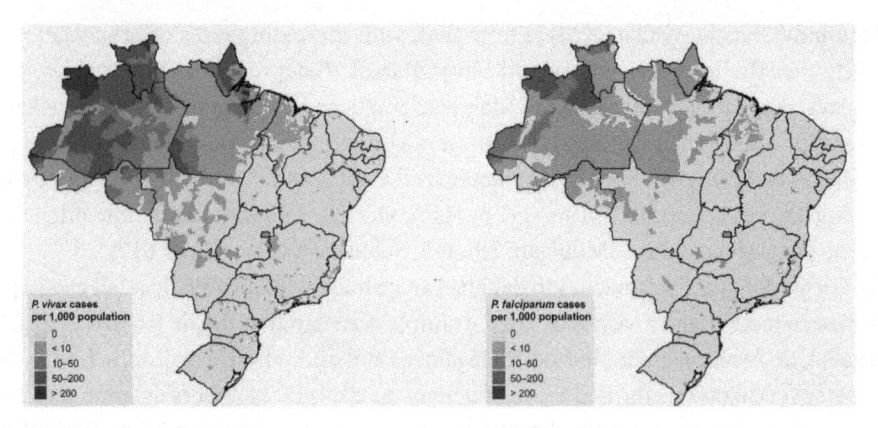

Figure 8.2 Spatial variation in malaria linked to two parasite species in Brazil, 2018
Source: Lana *et al.* (2021).

migration, both within the country and across international borders with Thailand and Cambodia (Kounnavong *et al.* 2017). Research suggests that this is true at the border between Thailand and Myanmar, where military conflict and movements of refugees (Chapter 5) add another layer of complication.

Ferring and Hausermann (2019) show how the environmental and social consequences of gold mining have shaped malaria incidence in central Ghana. Setting aside classical epidemiological explanations of such incidence that identify separate risk factors, as well as the tendency to blame individuals for not using protective bed nets, they argue that "mining's socioecological outcomes – from food insecurity and water-logged pits to mercury exposure – combine with social inequalities and *Plasmodium falciparum*'s [the malaria parasite] unique biological capabilities to render women and children most vulnerable to malaria" (Ferring & Hausermann 2019: 1077). Vulnerability accumulates, as a result of the intersection of environmental degradation, malnutrition, age and gender. Health in this region of the Global South involves more-than-human assemblages.

Networked disease I: Ebola

The outbreak of Ebola in west Africa in 2014–16 that killed over 11,000 people offers a compelling example of the important role played by geographical distance and networked space (but also structural inequalities) in the transmission of disease. The disease is thought to have been transmitted from the bite of a bat on a two-year-old boy living in Guinea (a so-called "spillover" event), but it spread quickly through geographical space.

High rates are found where the three most affected countries – Guinea, Liberia and Sierra Leone – meet (Figure 8.3). West Africa is characterized by a high degree

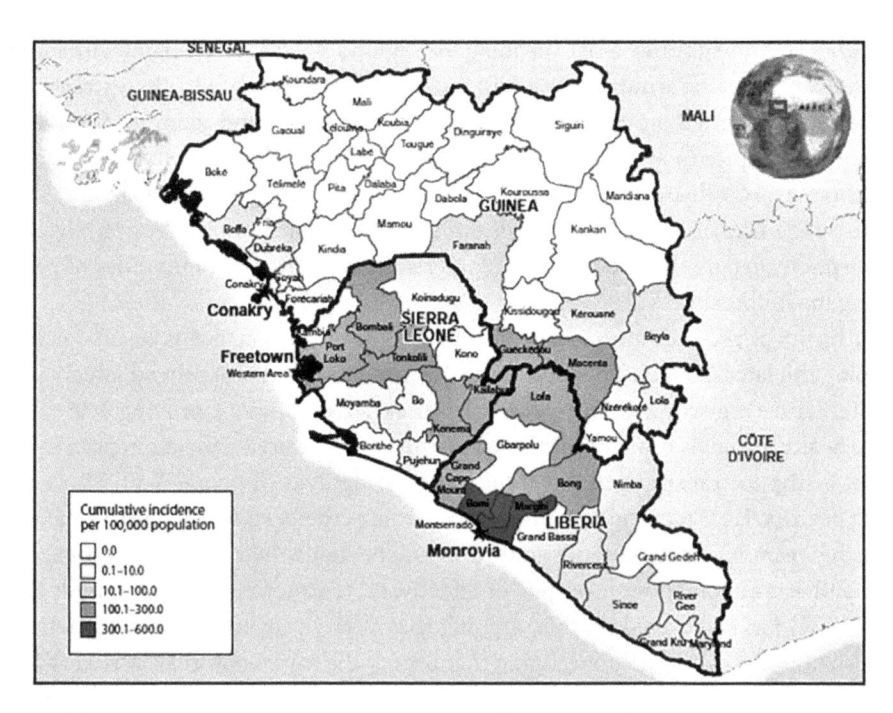

Figure 8.3 Ebola virus disease cumulative incidence, 20 September 2014
Source: CDC (2014).

of population movement, for work, food or shopping in markets, across exceptionally porous borders, such as between northwest Sierra Leone and Guinea. A nurse, Els Adams, working for Médecins Sans Frontières in Kambia, northwest Sierra Leone, reports that, "for the local population, the border hardly exists. There are at least 34 different crossing points, and probably even more. People identify with the region rather than the country" (Doctors Without Borders 2015). Language, local cultures and family relationships bind the three countries together.

Such mobility has consequences both for disease spread and control. Contact tracing becomes extremely challenging when people have traversed borders looking for hospital care. The custom of travelling sometimes considerable distances, again across borders, to return home to die and be buried near family adds another layer of risk to disease transmission.

Although these local transport networks are clear evidence of the proximate causes of disease spread, we must look elsewhere for the deeper structural causes. There is a history, as well as a geography, to Ebola in west Africa, both of which come through clearly in perhaps the most detailed "biography" of the disease (Farmer 2020). "Upstream" has a temporal as well as a spatial meaning: socio-spatial context matters, but so too does social history. For Farmer, the 2014–16 outbreak reflects the "ravages of history" (as his subtitle has it), meaning the region's "long entanglement with Europe and the Americas" (Farmer 2020: xxvi). Such connections relate to the extraction of wealth (latex from Liberia, bauxite from Guinea, diamonds and titanium from Sierra Leone: see Chapter 6) and from forests, which "have contributed acutely to the Ebola crisis" (xxiii). Such historical entanglements, of raw materials flowing from the Global South to the Global North ("diamonds are forever", as Farmer observes pithily: 358), are distinctly geographical but on a far larger scale than the local mobilities of people during the outbreak.

As for local geographies of care during the outbreak, Farmer is again scathing. Echoing the later Covid-19 pandemic, healthcare infrastructure proved largely inadequate, with care workers and personal protective equipment proving scarce. For those health agencies brought in to manage the Ebola outbreak, the main priority was to put a stop to transmission; he argues that saving lives of people with Ebola took lower priority. He is particularly critical of the role of the WHO, which was "weakened over the years by political feuds and funding shortfalls, had little operational capacity and was unable to act promptly or effectively" (Farmer 2020: 61). Kamradt-Scott (2016: 401) has elaborated on this, arguing that WHO proved "inept, dysfunctional, even shambolic", while acknowledging that during the initial outbreak period (March to June 2014) governments in all three affected countries sought to downplay the extent of the outbreak for fear of trade or travel sanctions and therefore loss of foreign exchange.

Drawing on interviews with healthcare staff, Hirsch (2021) examined the local geography of Ebola care in more detail, and argues that, whereas international and local healthcare workers alike were at risk of infection, the risk of death was greater among the latter (Black) group than the former (predominantly white), reflecting the differential access to healthcare facilities. The spatial organization of such facilities meant that some lives were considered "more worthy of different standards of care – through access to European-style treatment facilities and the possibility of being medevaced [evacuated by air] – than others" (Hirsch 2021: 6). Racial inequality reared its head in west Africa, as it had during colonial times. Although they do not comment on the racism embodied in differential access to care, Gee and Skovdal (2017) also note the separation of international and local care workers, identifying the former as occupying relatively privileged positions as they navigated their way through the Ebola "riskscape". Such a riskscape was constructed in the popular press as reflecting local culture (eating infected bushmeat, for example) and the inability of local populations to comply with experts in disease management. However, and echoing Farmer, some writers suggest that such a narrow viewpoint represents "a smokescreen that enables and perpetuates ongoing structural inequities – notably, by omitting consideration of global power relations, colonial history and contemporary extractive political economies" (Richardson, McGinnis & Frankfurter 2019: 1).

Networked disease II: influenza viruses and SARS

The emergence of Covid-19 has supplanted earlier concerns with new strains of influenza: avian influenza ("bird flu") and swine influenza. Indeed, writing in 2010 Herring and Lockerbie described avian influenza as *the* coming plague" (Herring & Lockerbie 2010: 179, emphasis added), suggesting that there is a "deep foreboding that a global cataclysm lurks in the farming communities and markets of Southeast Asia, waiting to be unleashed" (191). As it happens, they were quite correct – except that the cataclysm that began in December 2019 proved not to be an influenza virus. Indeed, avian influenza (the H5N1 virus) has, according to the WHO, spawned fewer than 900 cases and 500 deaths since 2003. The "cataclysm", such as it was, was the financial loss borne by the poultry industry, especially family farmers in poor rural communities whose flocks were culled.

Swine flu (H1N1) was first detected in 2009 in Mexico. As with other viruses, research has modelled global spread via airline transport, suggesting a close correlation between the destination countries and the known cases of H1N1 in such countries (Khan *et al.* 2009). The mortality burden from H1N1 is uncertain, but the US CDC suggests that, worldwide, up to half a million people may have died.

In an important paper, Sparke and Anguelov (2012) discuss the inequalities engendered by the 2009 H1N1 "pandemic" (as it became labelled by the WHO). Inequality proved both a determinant and a consequence of H1N1, a point that applies to all geographies of infectious (and other) disease. These inequalities, they suggest, took four strands: blame; risk management; access to therapies; and in how the virus originated. Its origins in Mexico led some commentators to label it "Mexican flu" (in much the same way as ex-President Trump referred to SARS-CoV-2 as the "Chinese" virus), thereby conferring stigma on Mexican immigrants. The fact that a Mexican boy (thought to be "patient zero") may have inhaled the virus from pigs in a US-owned pig-fattening factory went largely unremarked. The risks of H1N1, the authors argue, bore no comparison to the burden of illness and mortality generated from other diseases, such as malaria and childhood diarrhoea. Echoing later discussions of Covid-19, there were inequalities in potential access to flu vaccines. Last, and quoting the American commentator Laurie Garrett, they note that "the real place to blame is the strange ecology we have created to feed meat to our massive human population" (Sparke & Anguelov 2012: 733). Their discussion of the transformation of agriculture in southern China that has enabled it to become an epicentre for influenza viruses is a theme that bears closer examination with respect to SARS-CoV-2. They echo a quote in Herring and Lockerbie (2009: 190), namely that "small farms have outbreaks, big farms breed epidemics; globalization of big farms creates pandemics".

As the world continues to grapple with the coronavirus (SARS-CoV-2) that has caused Covid-19, it is instructive to return to a similar virus (SARS-CoV-1, usually abbreviated just to "SARS" – severe acute respiratory syndrome) that caused illness (but fewer than

10,000 known cases), death (but fewer than 1,000 cases) and economic devastation (financial losses of $54 billion globally) in 2003 (Ali & Keil 2008: 3). Other than these statistics, which are tiny in comparison with SARS-CoV-2, the similarities with the latter's disease origins, spread and management (quarantine, "social distancing" and hygiene protocols) are striking.

The first case of SARS is thought to have been a person from the city of Foshan in Guangdong province, China, who had apparently eaten civet (wild cat) bought in a meat market, although subsequent research has found the virus in horseshoe bats that passed the disease to civets (Cyranoski 2017). But it is the wet markets in the main provincial city, Guangdong, that are considered the breeding grounds for SARS. These markets (also found in other countries: Figure 8.4) have expanded to meet the demand for live poultry, fish and reptiles, as well as more exotic animals (such as the civet cats) that harbour the virus.

Having originated in Guangdong province, China, the disease was the archetypal signifier of globalization, spreading to Hong Kong, Vietnam, Singapore, and Canada (see Gatrell 2011: 143 for details of local and global spread). As an editorial in Toronto's main newspaper asserted, "Globalization means that if someone in China sneezes, someone in Toronto may one day catch a cold. Or something worse – if, in Guangdong province, 80 million people live cheek by jowl with chickens, pigs and ducks, so, in effect, do we all. Global village indeed" (van Wagner 2008: 13). The network of cities connected

Figure 8.4 Caged birds for sale in a wet market in Shanghai, China
Source: iStock.

by air traffic brought China close to Canada in time-space. Yet, as the Toronto editor recognized, the networks were not simply those facilitating the movement of people from one place to another: "There are significant human and non-human aspects of connectivity that are not adequately confronted by our images of network connectivity" (van Wagner 2008: 23). Different elements formed an assemblage that set both the foundations and consequences of SARS. Poor sanitation in the wet markets, the cramped conditions in which the animals were housed and the widespread use of antimicrobials all contributed to this assemblage. The consequences of the overuse of antimicrobials are discussed in Box 8.3.

BOX 8.3 ANTIMICROBIAL RESISTANCE

According to the WHO, about 10 million people across the globe fell ill with tuberculosis in 2020, and 1.5 million died (many also with HIV). The bacterium is airborne and therefore can be spread from person to person through close contact. The disease is spread by crowding people into small spaces. It is treatable with a range of antibiotics (or antimicrobials) but drug-resistant strains have appeared, rendering some of the "first-line" drugs ineffective; hence, multi-drug-resistant tuberculosis (MDRTB) and extensively drug-resistant tuberculosis (XDR-TB) have become, as far as the WHO is concerned, security threats (see Chapter 4).

Such resistance is not only a feature of TB. Others working in public health consider the more widespread emergence of antimicrobial resistance to be a major public health crisis, both in the Global North and Global South. England's former chief medical officer, Sally Davies, describes it as a "silent pandemic" and an "existential threat to modern medicine" (Davies 2021). In part, this is because of overuse or misuse of antimicrobials, both in human and animal populations. Since 2000 the use of antibiotics has grown considerably, particularly in low- and middle-income countries. However, in these countries antibiotic treatments are needed to manage the burden of infectious disease.

Many common pathogens are becoming resistant to treatment; they include *Campylobacter*, *Salmonella* and MRSA (methicillin-resistant *Staphylococcus aureus*). Frost et al. (2019) document the role of international travel in diffusing these drug-resistant pathogens across the globe. People (including medical "tourists") travelling to regions with high rates of AMR prevalence may be exposed to resistant bacteria and return home with these infections. Improvements to sanitation and water quality – as well as nutrition – will reduce the burden of diarrhoeal disease and hence the need for antibiotics; attention needs to be focused on the "upstream" causes of AMR in LMICs.

There is also a wider ecological concern, since antibiotic-resistant bacteria linked to farm animals (kept in cramped conditions) can be transmitted to human populations via food chains, and spread via animal wastes. One example is the overuse of antibiotics (such as colistin) to promote the growth of pigs in China (now banned). The geographical spread of a gene (*mcr-1*) that confers resistance to colistin has been enabled by the transport of animals and meat products.

Further online information on antibiotic use and resistance may be found at www.resistancemap.cddep.org, and see www.amrleaders.org/about-us for progress on international action to tackle the issue. Cliff *et al.* (2009: 221–38) have a good geographical description of the issue, while Hinchliffe (2022) offers an alternative perspective on AMR, criticizing its emphasis on surveillance and the focus on behaviour change. He suggests, for example, that replacing small-scale poultry farms with large-scale industrialized farming is hardly conducive to disease control. "Modernization under increasing climate stress can in this case produce the conditions for poorer animal health outcomes and, as a result, increased pressure to utilize antibiotics" (Hinchliffe 2022: 158).

The unequal geographies of Covid-19: pandemic or syndemic?

Any treatment of the geographies of Covid-19 requires more than the section of a chapter – indeed, more than a chapter. Unsurprisingly, a number of books have been, and will continue to be, published on Covid-19, including several that draw explicitly, or implicitly, on a geographical imagination (Christakis 2020; Andrews *et al.* 2021). Here, I consider three key themes: the geographical origins and spread of the SARS-CoV-2 virus; variations in incidence and mortality, with particular reference to the Global South; and variations in the availability of vaccines to prevent infection. It will come as no surprise that inequality is the thread that runs through all three themes.

Although most scientists believe that SARS-CoV-2 originated in animals (horseshoe bats, for example) and crossed the species barrier into human populations (either directly, through bites, or via an intermediate animal in wet markets), the notion that the virus escaped from a virology laboratory in Wuhan, China (the centre of the outbreak), continues to have traction; Worobey *et al.* (2022) provide strong evidence of clustering around the Huanan seafood market in Wuhan. As we have seen with other viruses, we cannot hope to understand the virus without acknowledging the links with what Sparke and Anguelov (2020: 499) call the "neoliberalization of nature (including the wholesale privatization, marketization and financialization of the natural world)". Such a perspective was prefigured by Davis' (2006) reference to the "monster at our door" and Wallace (2016), who puts agro-industry centre stage, although both were more concerned with influenza viruses. Gibb *et al.* (2020a) back up these bold assertions with research suggesting that land use change brings people into contact with livestock and reservoirs of zoonotic disease. The expansion of industrialized agriculture, and urbanization in low- and middle-income countries, threaten to create further spatial settings for human exposure to zoonotic pathogens.

If the origins of SARS-CoV-2 lie in human interference in ecosystems in parts of the Global South, the spread of the infection reveals further geographical connections. Given that the pandemic originated in the city of Wuhan, in Hubei province, China, it is clearly of interest to understand how the virus diffused spatially from that source. Jia *et al.* (2020) have undertaken just such a study, using data on almost 12 million mobile phone calls to track population movement during January 2020 (until quarantine was introduced) from Wuhan to almost 300 destinations in China. Their analysis reveals clear evidence of contagious spatial diffusion (that is, a function of proximity to Wuhan), with additional evidence of hierarchical diffusion as people travelled to other major cities (notably to Beijing in the northeast and Hong Kong to the south); for confirmation of such diffusion from the "hearth" (Hubei province, the capital of which is Wuhan), Figure 8.5 is illuminating.

By January 2023 the number of cases of Covid-19 had reached over 670 million, with about 6.7 million deaths (Our World in Data). However, in an important paper that reviewed Covid-19 deaths over two calendar years (2020 and 2021), the COVID-19 Excess Mortality Collaborators (2022) have suggested that the number of deaths

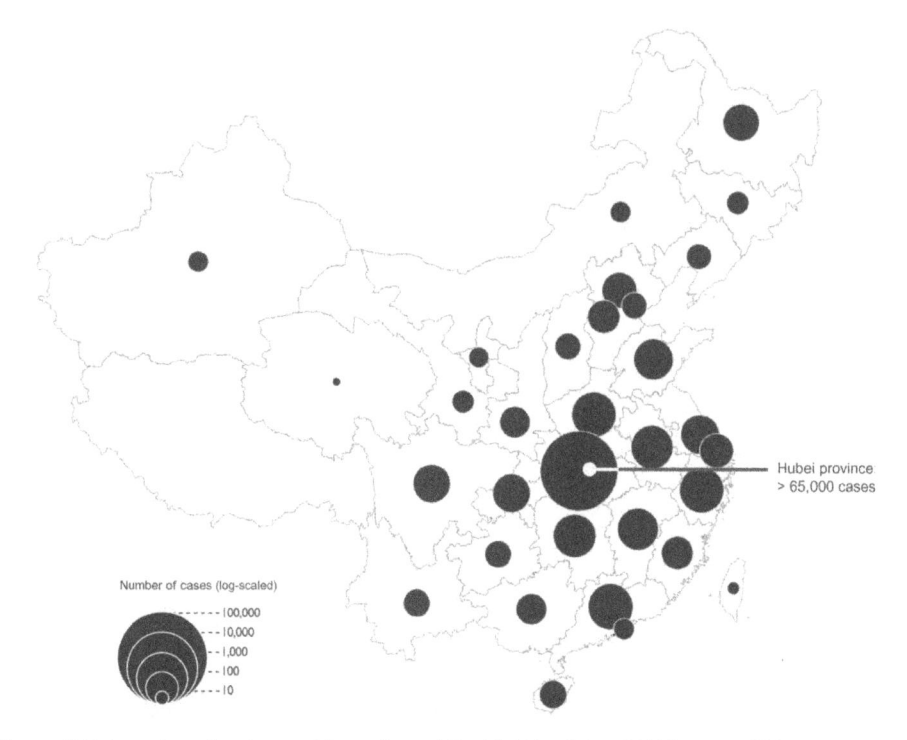

Figure 8.5 Location of patients with confirmed Covid-19 in China, 19 February 2020
Source: Esri (2020) (courtesy of Ken Field).

worldwide because of the pandemic (excess mortality) is likely to be three times the number of reported deaths. Their paper estimates excess mortality for 74 countries as well as subnational estimates for several countries, including Mexico, India and Brazil (Table 8.2). As is clear from their analysis, broad national figures mask considerable regional variation. For example, excess mortality rates in western Brazil (Amazonas and Rôndonia states) are well above the national average, as is Ceará in the northeast; however, both Alagoas and Paraiba, also in the northeast, have relatively low rates of excess mortality. Early on in the pandemic Brazil's then president, Jair Bolsonaro, took pride in dismissing concerns about the virus, joking that during his New Year's swim in 2021 "I dived in with a mask on so I wouldn't catch COVID from the little fish" (Ottawa CityNews 2021).

To add a different perspective on the scale of the pandemic, Table 8.2 also shows the absolute numbers of reported Covid-19 deaths, revealing the huge toll in Brazil, India and Mexico, as well as that in just one Indian state, Maharashtra, where Mumbai, its major city, will have accounted for the majority of deaths. Whether these reported deaths prove to be grossly inaccurate remains to be seen. It has been suggested that the number of Covid-19 deaths in India is closer to 4 million, not the roughly 500,000 acknowledged by the Indian government (Ellis-Petersen 2022).

Table 8.2 Estimated excess mortality rate (per 100,000), and reported Covid-19 deaths, by country/state, 2020/21

Country/region		Estimated excess mortality rate (per 100,000)	Reported deaths
Global		120.3	5,940,000
Bolivia		734.9	19,700
Eswatini		634.9	1,300
Lesotho		562.9	665
Peru		528.6	203,000
Lebanon		416.2	9,120
Botswana		399.5	2,440
Namibia		395.6	3,630
Ecuador		333.4	34,100
Mexico		325.1	418,000
Brazil		186.9	619,000
	Rôndonia	269.0	6,730
	Ceará	255.1	24,800
	Amazonas	246.1	13,800
	Paraiba	107.6	9,600
	Alagoas	97.8	6,380
India		152.5	489,000
	Uttarakhand	284.6	7,420
	Manipur	263.8	2,000
	Maharashtra	259.5	142,000
	Arunachal Pradesh	56.4	282

Source: COVID-19 Excess Mortality Collaborators (2022).

Covid-19 bears all the hallmarks of a syndemic (or, indeed, an ecosyndemic: Box 8.1). It "is not simply a viral infection, but rather a complex syndemic with clinical and structural vulnerabilities entrenched by existing poor health, employment and precarity, deprivation and inequalities" (Cousins *et al.* 2021). The clear association between excess deaths because of Covid-19 and income inequality across many countries has been revealed by Varkey, Kandpal and Neelsen (2022).

Evidence suggests considerable geographical overlap between Covid-19 infections and the incidence of non-communicable disease, while people living with HIV have seen an increased risk of hospitalization. There is also increasing evidence of links between exposure to air pollution and susceptibility to hospital admissions from Covid-19 (Stewart 2021). Conversely, the widespread adoption of social distancing or "lockdown" measures, restricting vehicular movement, has led to unanticipated benefits, such as reductions in air pollution in many of the world's most polluted cities. However, these improvements seem to have been short-lived. According to the European Space Agency (ESA), although nitrogen dioxide concentrations in Beijing fell by over a third between February 2019 and 2020, by February 2021 they had returned to pre-pandemic levels (ESA 2021).

It has been established in many studies undertaken in the Global North that there are clear, and preventable, inequalities in Covid-19 incidence and mortality. For example, according to the UK Office of National Statistics, in England both men and women in the Bangladeshi, Pakistani, Black African and Black Caribbean ethnic groups have had an elevated risk of mortality compared with white British. The same is true of Native Americans in the United States; Wang (2021) describes the disproportionate impact on the Navajo nation.

Similar inequities exist among Indigenous peoples in the Global South. In Brazil, for example, Soares *et al.* (2022) have shown that Indigenous people have suffered disproportionately in comparison with non-Indigenous Brazilians. In 2020 there was, overall, for the former group a 35 per cent increase in mortality compared with what was expected, and an 18 per cent increase in the non-Indigenous population. Along the Amazon the virus spread from Manaus to isolated towns and villages on crowded boats.

These disparities are mirrored for Indigenous peoples in Colombia, Ecuador and Mexico (Soares *et al.* 2022). In Mexico there was an elevated mortality risk among those diagnosed with Covid-19 who were living in municipalities where income levels were low, housing conditions were crowded and there were higher proportions of Indigenous people (Ríos, Denova-Gutiérrez & Barquera 2022). All these markers of poverty and vulnerability are likely to be associated with higher risk of disease transmission.

In a powerful paper, Büyüm *et al.* (2020) have linked the unequal impact of Covid-19 to structural violence wreaked on marginalized communities. For example, they describe the overcrowded conditions in which Singapore's migrant workers live, noting that, although Singapore has had some success in managing the pandemic, "SARS-CoV-2 has lifted the smokescreen to reveal how little these workers are actually valued, resulting in Singapore's failure to protect them from the virus and to protect the entire nation from a resurgence in cases" (Büyüm *et al.* 2020: 2). Such overcrowding – resulting in the impossibility of maintaining social distance – is clearly a factor contributing to elevated disease incidence, and mortality in other places, whether refugee camps, urban squatter settlements or prisons.

As we saw with Ebola, it is impossible to halt the spread of Covid-19 across land borders. A good example of the issue comes from looking at the spread of the SARS-CoV-2 virus from Brazil to Uruguay. Compared with some countries in South America, Uruguay had some early success in managing the pandemic, but sharing a 1,000 km border with Brazil that is characterized by considerable cross-border movement meant that the virus diffused spatially into Uruguay (Mir *et al.* 2021).

Government responses to the pandemic, whether the roll-out of vaccination or non-pharmaceutical interventions such as quarantine (lockdown) and social distancing, have varied from country to country. As Cousins *et al.* (2021: 3) have noted, "The uneven impact of the coronavirus pandemic has been a reminder of the difference between contexts where strong public health systems exist (e.g. New Zealand, Germany, Vietnam) and those where public health has been eroded by austerity and neoliberal approaches to healthcare (e.g. UK, USA, Brazil)." It is easy to point a finger of blame against so-called "super-spreaders" of the infection while forgetting that some policies of national governments meant that the latter were the real super-spreaders.

I acknowledge the efforts made to develop and produce vaccines against Covid-19, but the availability of, and access to, such vaccines has exposed further global inequalities (Yamey *et al.* 2022). Data (Table 8.3) reveal which countries had, in April 2022, fewer than 5 per cent of their population fully vaccinated, while a global map conveys inequalities in gross numbers vaccinated the previous year (Figure 8.6).

It is far too simplistic to claim that those in the Global North acted more quickly or efficiently than those in the Global South. Büyüm *et al.* (2020) report that Senegal helped reach vulnerable entire populations with affordable tests for the virus, but that

Table 8.3 Countries with fewer than 5 per cent of the population fully vaccinated, April 2022

Country	Percentage fully vaccinated
Burundi	0.09
Haiti	1.02
Papua New Guinea	2.93
Madagascar	3.90
Cameroon	4.05
South Sudan	4.36
Malawi	4.63
Chad	4.74
Nigeria	4.76
Mali	4.98

Source: https://coronavirus.jhu.edu.

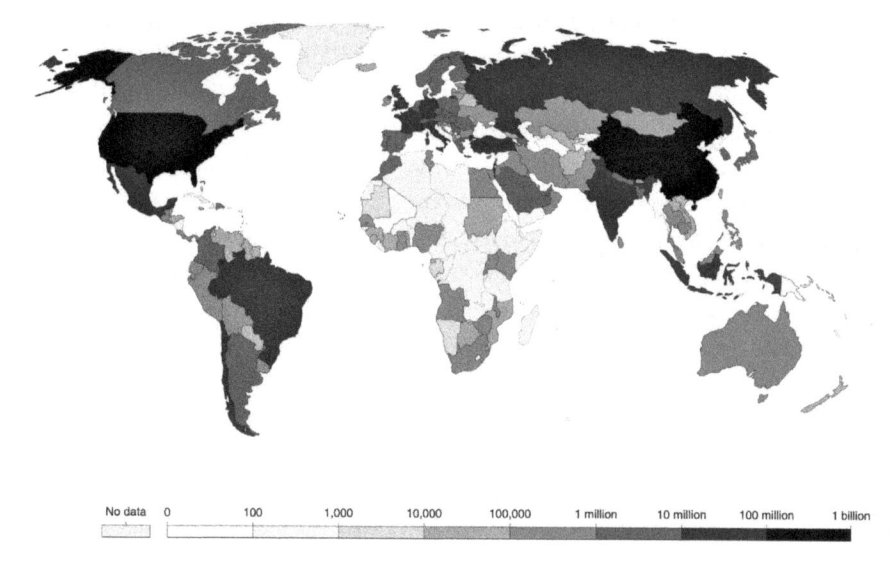

Figure 8.6 Global inequality in number of Covid-19 vaccine doses administered, April 2021
Source: www.nature.com/articles/s41562-021-01122-8/figures/1.

"international coverage of the continent … focused on the assumed inevitable failure of African nations to effectively respond to the pandemic, failures which are often caused by limited resources resulting from colonialism and modern-day imperialism".

Kerala, in southwest India, implemented well-coordinated lockdowns and testing and contact-tracing strategies to contain viral spread, as reported by those working there in public health (Menon *et al.* 2020). Büyüm *et al.* (2020) note that such early successes were countered by widespread negative media coverage of the country as a whole. But this was very much an interim statement, and, as with so much relating to Covid-19, it may be several years before a complete picture emerges of the pandemic and accompanying data.

As vaccines started to appear in early 2021 pharmaceutical companies were accused of protecting their patents, and governments in wealthier countries reportedly engaged in a "frenzy of deals" to buy up large stocks of vaccines, leaving poorer countries with fewer supplies (Sparke & Williams 2022). This became known as "vaccine nationalism". In contrast, some countries have engaged in "vaccine diplomacy", a form of soft power to improve or influence relations between a donor and recipient country. China has donated substantial quantities to countries in Africa, but is far from alone in engaging in such activity. Its vaccines (mostly Sinovac and Sinopharm) are donated to 49 and 90 countries respectively, in many parts of Africa and South America, while the Russian Gamaleya (Sputnik V) vaccine has been distributed in India and north Africa (*New York Times* 2021). The major company Pfizer offered in May 2022 to make available to 45 low-income countries all current and future vaccines on a not-for-profit basis (*BMJ* 2022).

COVAX (Covid-19 Vaccines Global Access) is a global initiative, involving several partners (GAVI [the Vaccine Alliance, originally the Global Alliance for Vaccines and Immunization]; the Coalition for Epidemic Preparedness Innovations [CEPI]; and the World Health Organization). It aims to provide equitable access to Covid-19 vaccines via donations from wealthy countries and vaccine manufacturers. Although some such countries may have seen a moral case to assist resource-poor countries, they also recognized that the global economy could lose as much as $9.2 trillion (because of lost trade and the potential spread of new variants of the virus) unless poorer countries had adequate access to vaccines (ICC [International Chamber of Commerce] Research Foundation 2021).

Puyvallée and Storeng (2022) have cast a critical eye on COVAX, suggesting that political (self-)interest among potential donor countries has limited its ability to implement effective dose-sharing, and hence to address global inequities. Some countries preferred to donate directly to specific countries (notably, former colonies) rather than taking part in COVAX. There have been time lags between offers to donate and vaccines being delivered, and the short shelf life of some vaccines proved unhelpful. Echoing our earlier discussion of security (Chapter 4), the authors argue that donations have sought to protect national security interests.

Israel makes for an interesting story, not least because of the contrast with the neighbouring West Bank territory of Palestine. By late February 2021 Israel had vaccinated more than 30 per cent of its population with two doses while in Palestine the figure was under 1 per cent. The charity Médecins Sans Frontières put it quite starkly in 2021, saying that "you are over 60 times more likely to have a vaccination in Israel than in Palestine" (Kennes 2021). By April 2022 about 40 per cent of the eligible Palestinian population had received at least one dose, compared with 74 per cent in Israel (*New York Times* 2021).

Concluding remarks

A chapter on infectious disease brings together all the key concepts and perspectives that are essential to the study of global health: political ecology, mobilities, inequality, security, globalization and time–space convergence. Cousins *et al.* (2021) point to the intersection of the three crises that have emerged in the 2020s: the Covid-19 pandemic, climate change and inequality. New syndemics have arisen, bringing together vulnerabilities resulting from malnutrition (worsened by the war in Ukraine), climate and land use change and the struggles to cope with both non-communicable disease and infections other than from coronaviruses. Inevitably, these vulnerabilities are socially and politically determined. Time will tell what other infections emerge to impact on global health.

Globalization was defined earlier (pp. 10–12) as the intensification of worldwide social relations that link distant localities in such a way that local happenings are shaped by events occurring many miles away, and vice versa. As we have seen in this chapter, this applies very much to the geographic spread of many infectious diseases.

What started in "distant localities" in China, whether SARS or Covid-19, has diffused rapidly and widely to places in both the Global North and Global South. Governments in the Global North have been quick to construct such spread as threats to security, and give much less attention to those diseases that have affected, and killed, many more people in recent years.

Further reading

For an excellent, and highly readable, consideration of neglected tropical diseases, the best starting point is Hotez (2008). Medical geographers Peter Haggett, Andy Cliff and Matthew Smallman-Raynor have written extensively on the spatial analysis and modelling of infectious disease, including some diseases not considered in this chapter; see, for example, Cliff *et al.* (2009).

On the geographies of infectious disease, I recommend the collection edited by Herring and Swedlund (2010), and particularly, with regard to SARS, the chapters in Ali and Keil (2008). Readers may find Gatrell (2011), especially chapters 8 and 9 on disease spread, worth consulting. Bashford (2006) has several chapters worth reading, including on SARS. For a recent discussion of syndemics and ecosyndemics, see Tallmann *et al.* (2022).

A very good overview of migration and is links to population movement is provided by Martens and Hall (2000). You should certainly examine the WorldPop website (www.worldpop.org/about) to search for available data, as well as references to the prolific research of that group.

On Covid-19, see Boyle (2021), who has a final chapter, "Coda on COVID-19", that is worth reading. More generally, see Spiegelhalter & Masters (2021) and Sridhar (2022). The recent essay by Cousins *et al.* (2021) and the papers by Sparke and Williams (2022) and Herrick, Kelly and Soulard (2022) are all deserving of a careful read. Anderson (2021) offers a thoughtful critique of epidemiological modelling of Covid-19. Utazi *et al.* (2022) have explored variations in vaccination rates in LMICs. The *Lancet* Covid-19 Commission (Sacks *et al.* 2022) on "lessons for the future" is an important paper.

Amzat and Razum (2022: chs 6 & 7) deal with infectious disease and Covid-19.

9
CLIMATE CHANGE AND GLOBAL HEALTH

> Global warming is largely a by-product of social structures and patterns, namely the self-serving belief held by global elites and powerful corporations that the world has unlimited resources and the wedding of that belief to a neoliberal governmental policy commitment to the unrestrained flow of capital, ever-increasing commodity production, and the promotion of consumption-based lifestyles.
> Singer (2010: 32)

Climate change is but one component of a set of wide-ranging human impacts on the environment, which also include urbanization, land use change and loss of biodiversity, deforestation, resource extraction on land and in the oceans, the growth of agri-business, and conflict. It must therefore be borne in mind that climate change is part of a complex system of environmental disturbance and perturbation. The introductory quote by Merrill Singer makes it quite clear how embedded is the deeper structure of this complex system. The present chapter seeks to show how the climate crisis is also a health crisis.

There is abundant evidence of long-term climate change. For example, the United Kingdom's Meteorological Office reported in May 2022 that climate change has increased by 100-fold the chances of record-breaking heatwaves in northwest India and Pakistan; a heatwave (of more than 50°C) is projected to occur every three years (Met Office 2022). There was a fivefold increase in exposure to extreme heat events between 1980 and 2017 for the world's most populated cities, and current exposure to heatwave events is almost 15 billion person-days per year, with the greatest cumulative exposures occurring in southern Asia (7.19 billion), sub-Saharan Africa, and north Africa and the Middle East (Met Office 2022).

Any discussion of the impact of climate change on human health must begin with the monumental report from Working Group 2 of the Intergovernmental Panel on Climate Change (IPCC), which was delivered in February 2022 (IPCC 2022). At 3,675 pages this report is truly encyclopaedic, but the "Summary for policymakers" is an excellent starting point. My own focus is initially on impacts on vector-borne infectious disease (malaria and dengue), waterborne disease, nutritional deficiencies and food security. I then consider impacts on Indigenous groups before looking at the consequences for population displacement and migration. The latter, as with so much else on climate change, makes highly visible the "geographical connections" in this book title. Even so, I can only scratch the surface, and my focus is almost exclusively on the Global South and how climate change risks and impacts are unevenly distributed spatially and socially.

Inequality is a golden thread that runs through what follows. A dramatic illustration of this is when we look at international variations in carbon footprints (Bruckner *et al.* 2022). Many countries in sub-Saharan Africa, such as Burkina Faso, Madagascar and Rwanda, had estimated footprints in 2014 of less than $0.2tCO_2$, while that for the United States was $14.5tCO_2$.

For some writers, such as Sultana (2022), inequality reflects historical and contemporary coloniality (Box 9.1).

BOX 9.1 THE COLONIALITY OF CLIMATE CHANGE

The summary of the latest IPCC report has the following statement, which it offers "with high confidence". "Vulnerability of ecosystems and people to climate change differs substantially among and within regions, driven by patterns of intersecting socioeconomic development, unsustainable ocean and land use, inequity, marginalization, *historical and ongoing patterns of inequity such as colonialism*, and governance" (IPCC 2022: 16, emphasis added).

Sultana (2022) expands on this in a powerful paper, arguing that, since colonialism provided resources extracted from the Global South for the benefit of the Global North, countries in the Global South have been left with less capacity to address climate impacts. In the contemporary world economy "wealthier colonial and imperial states can export and offshore their emissions, while weaker countries or those with lax environmental regulations, have their emissions go up. It is a form of carbon colonialism." She argues that decolonizing climate would require wholesale restructuring of the global economy. Echoing Singer's introductory quote, she suggests that climate "coloniality" takes various forms, not only through "hyper-consumptive and wasteful lifestyles" but also via the deeper structures "shaped by capitalism and its powerful global alliances".

Heat-related mortality

There is a wealth of evidence relating human mortality to temperature (Gatrell & Elliott 2015: ch. 12). Much excess mortality is attributable to excessively low temperatures, but my emphasis here is on the impact of extreme heat on vulnerable populations, particularly those living in cities, where the urban heat island effect is pronounced. In a classic study, Klinenberg (2002) documents the impact on morbidity and mortality of very high July temperatures in Chicago in 1995; such impacts were socially patterned, with the risk borne disproportionately among those lacking air-conditioning who were on low incomes, were older people and were African-Americans. As he puts it: "Similarities between maps of the heat wave's deaths and maps of ethno-racial and class division reveal the social and structural underpinnings of the event" (cited in Gatrell & Elliott 2015: 160).

Here, I wish to consider some research on the impacts on populations living in the Global South. For example, in Latin America there are projections of considerable increases in mean annual temperature and in extreme heat events. Increased urbanization and a higher proportion of the population in older age groups mean that research is needed to model likely impacts on mortality. Kephart *et al.* (2022) looked at the relationship between mortality (15.5 million deaths between 2002 and 2015) and temperature across 326 cities. They find that small increases in temperature led to steep increases in cardiovascular and respiratory mortality among older adults. This was spatially variable, with cities in northern Argentina, southern Brazil and coastal Mexico particularly affected (Figure 9.1).

Climate change and infectious disease

Climate change means that many animal species will be forced to move into new areas to find suitable habitats, bringing pathogens with them and heightening the risk of zoonotic spillover. Research (Carlson *et al.* 2022) suggests that at least 10,000 different viruses capable of infecting humans may currently be circulating in animal populations (particularly bats). But the movement of mammals, triggered by climate change, means that a virus hosted by one (for example, bats) can get transmitted to another (so-called "novel virus sharing"), with risks of transmission to human populations. The authors write: "We predict that tropical hotspots of novel viral sharing will broadly coincide with high population density areas in 2070, especially in the Sahel, the Ethiopian highlands and the Rift Valley, India, eastern China, Indonesia, and the Philippines" (Carlson *et al.* 2022: 560). It would appear that, as well as being burdened by extreme poverty, large areas in the Global South are likely to bear the brunt of novel viral sharing.

Mora *et al.* (2022) have undertaken a comprehensive review of infectious diseases likely to be exacerbated by the hazards arising from climate change. For example, they suggest that global warming (leading to increased precipitation, flooding and drought) will lead to increased prevalence of over 100 infections. Most of these are vector-borne; others are waterborne, airborne or transmitted by direct contact or contaminated food. More specifically, I noted in the previous chapter that malaria is caused by *Anopheles* mosquitoes carrying the *Plasmodium* parasite. The ranges of the mosquitoes depend on temperature and rainfall, with an optimal temperature of 25°C. The relation between

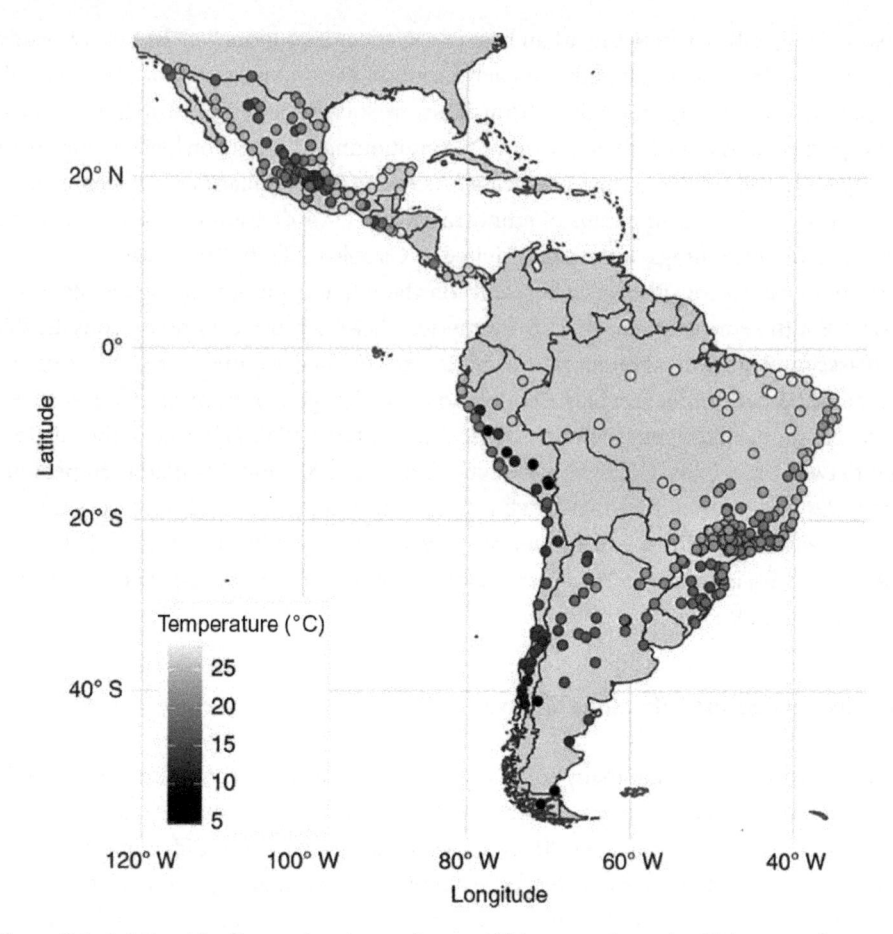

Figure 9.1 Relative risk of heat-related mortality per 1°C increase above the 95th percentile observed daily temperature
Source: Kephart et al. (2022).

disease transmission and temperature is non-linear; if the temperature is too high or too low, neither mosquito nor parasite will survive. Increased rainfall may mean more breeding sites, whereas higher temperatures may lead to uncovered water containers.

If we want to establish which regions – and, indeed, local communities – are at increased risk of malaria, we need detailed spatial climate data, as well as ancillary data that describe habitat suitability for the *Anopheles* mosquitoes and detailed population data to quantify risk. Among various examples of this kind of "What if?" modelling, Ryan, Lippi and Zermoglio (2020) have produced estimates of possible exposure to malaria for Africa (up to 2080). Although some parts of sub-Saharan Africa may see increased risk, others may not. However, in general, large parts of eastern Africa (Kenya, Tanzania, Uganda) and southern Africa (Angola, Namibia) could see over 75 million people at increased risk of the disease because of climate change.

The same considerations apply to the *Aedes* mosquitoes that transit the dengue virus. The IPCC report suggests that parts of Asia will be at serious risk from the dengue virus, but, as with malaria, this varies regionally, as well as seasonally (with higher incidence during the monsoons). Research in India (Kakarla *et al.* 2020) used a climate prediction model to predict the likely geographical distribution of dengue. Results suggest increasing incidence in most parts of India (which the authors describe as the "global epicentre" of dengue), apart from in the north. Elsewhere, Mordecai *et al.* (2020) suggest that where temperatures are regularly between 25 and 29°C, including many parts of sub-Saharan Africa, a warmer climate will suit dengue, chikungunya and other viruses transmitted by *Aedes aegypti*. In particular, they suggest that towards the middle and latter half of the century the likely hotspots of these viruses will shift east, away from the current focus in west Africa (Figure 9.2), although this is based only on temperature increases and is a scenario, not a prediction.

Increased temperature and rainfall are not the only factors that will lead to a widening of the range of the mosquitoes. Population movement and urbanization, and further human modification of the environment, also play a significant part. For example, drought means water needs to be stored, and, if such storage is in open containers, these will offer opportunities for mosquitoes to lay eggs. Deforestation and other human encroachment into wildlife habitats increase exposure risk. Gibb *et al.* (2020b: 4) express this clearly: "Models that integrate ecological or biological knowledge of important reservoir or vector species with near real-time climate and earth observation data can inform forecasts of certain zoonotic hazards weeks or months in advance." As with so much in relation to vector-borne illness, a disease ecology approach is required.

This is also true of waterborne infections. The parasite that causes schistosomiasis and the snails that host the parasite are both temperature-dependent, so there is potential for the range of the disease to spread. McCreesh, Nikulin and Booth (2015) have produced detailed maps to show where, in eastern Africa, changes in temperature may lead to the infection (*Schistosoma mansoni*) spreading to new areas. But increased risk

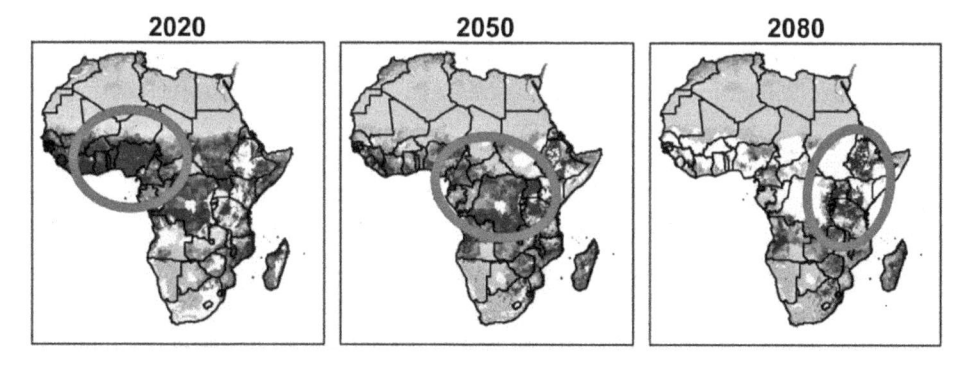

Figure 9.2 Possible shift of infections transmitted by *Aedes aegypti* mosquitoes from 2020 to 2050 to 2080
Source: Mordecai et al. (2020).

is not spatially uniform: some areas are predicted to see reduced risk. Countries such as Burundi and Rwanda, and eastern Zambia and southwest Kenya, are all likely to be at risk; in contrast, other parts of Kenya, as well as the east of the Democratic Republic of the Congo, may see reduced risk. The study is only suggestive of risks because of temperature increases and, as the authors note, other climate factors (notably rainfall) and determinants such as land use modification, population mobility and improvements in sanitation will all affect infection risk.

Drought and flooding can both have serious impacts on the spread of infectious disease, notably cholera, as they affect water use, sanitation and hygiene ("WaSH"), as Jones *et al.* (2020) have reviewed. A lack of clean water means that handwashing becomes impossible, while latrines get flooded when water levels rise. For example, recent cyclones in Bangladesh have destroyed riverside toilets, as the charity WaterAid has documented (WaterAid 2020), leading to more people having to resort to open defecation.

Climate change, food security and nutrition

Hunger affects 10 per cent of the global population. Its primary cause is poverty, but conflict lessens access to food supplies and the Covid-19 pandemic has worsened hunger, with the UN estimating that, because of the pandemic, an additional 118 million people were undernourished in 2020 compared with 2019 (WEF 2021).

The annual World Food Summit defines food security as being met when "all people, at all times, have physical and economic access to sufficient, safe, and nutritious food to meet their dietary needs and food preferences for an active and healthy life" (World Bank n.d.b). Key questions are how such access varies from place to place and among different groups, and how this will alter with climate change. I acknowledge the severe challenges placed on those on low incomes in countries in the Global North, but my focus is firmly on those living in the Global South.

SDG 2 aims by 2030 to "end hunger, achieve food security and improved nutrition and promote sustainable agriculture". This will prove hopelessly unrealistic. According to the Global Hunger Index in 2022, there are 47 countries at risk. Hunger is described as "extremely alarming" in Somalia, "alarming" in Burundi, the Central African Republic, Chad, Comoros, the DRC, Madagascar, South Sudan, Syria and Yemen. In 37 countries the situation is described as "serious" (von Grebmer *et al.* 2022). Since almost all these are in Africa, I focus my attention here on the impacts of climate change on food systems in that continent, drawing on some of the wealth of evidence in chapter 9 of the IPCC report.

Research shows that yields of crops such as wheat and maize are likely to decline as temperature increases by 1 to 2°C in tropical areas. Temperature affects yields, and these in turn affect food security, not least because of price increases, which affect access to food by poorer households. Further, the nutritional content of some of these crops may diminish. Raised levels of carbon dioxide may reduce protein concentrations of wheat, barley and rice by 10 to 15 per cent, contributing to child undernutrition (Swinburn *et al.* 2019: 801).

There are many communities in Africa at risk, whether farming crops or looking after livestock. Women engaged in agriculture bear a significant burden. Many have been displaced from rural to urban settings (see below, pp. 184–6). The IPCC report suggests "with very high confidence" that, between 2015 and 2019, over 45 million people in the Horn of Africa and 62 million people in eastern and southern Africa needed humanitarian assistance because of food emergencies prompted by climate change; children and pregnant women experienced the worst impacts on nutrition and health (IPCC 2022: 1350).

Increases in crop pests and associated diseases are predicted, evidence suggesting that they are contributing up to 35 per cent yield losses for wheat, rice, maize, potato and soybean in sub-Saharan Africa. Recent outbreaks of locust infestation in east Africa have been associated with climate change caused by ocean warming (Salih *et al.* 2020). With warming of 1.8 to 1.9°C, suitable areas for tea growing in Kenya and Uganda are projected to diminish by between 27 and 50 per cent. Variability in rainfall affects the availability of fodder for cattle, which are subject to malnutrition and reduced milk output. Climate change will threaten water availability for such livestock; droughts have become more frequent and widespread since the 1970s, and climate modelling suggests an increase in drought intensity and frequency for most of the continent if the climate warms more than 2.5°C.

There are predicted impacts on fish stocks. As sea temperatures in the tropics increase, some species of fish are predicted to move pole-wards, depriving some people in the Global South of food and livelihoods. Fish are an important source of protein, comprising almost two-thirds of the daily animal protein intake in parts of west Africa, such as The Gambia, Ghana and Sierra Leone (Allison *et al.* 2009). Regions thought to be most vulnerable to climate-induced changes in fisheries are in west and central Africa, the northwest of South America and parts of Asia (Pakistan, Bangladesh and Cambodia). In some areas, sea level rise risks the incursion of saline waters into inland rivers.

Empirical research such as that reported here *describes* the possible consequences, for food security, of climate change. But to *understand* such consequences we need to go beyond maps, graphs and statistics to hear the ongoing, and likely, impacts on individuals. Otherwise, research on, say, drought as a marker of climate change becomes itself very dry.

Nyantakyi-Frimpong (2021) has undertaken just such important work, looking at the impacts of climate change on women in Ghana's Upper-West region who have to juggle working on the land with infant care. Drawing on political ecology, he examines how the experience of climate change is embodied. One mother with young twins told him:

> I'm emotionally stressed about whether the changing climatic conditions will prevent successful agricultural yields, which will also worsen my child's nutrition. Am I going to replant again? I'm afraid and worried about juggling farm work and caring for my twins, who are less than 12 months. Think about how that affects my blood pressure and how that affects my peace of mind to care for the infants. I don't get time for them, and it makes me sad. I feel like a horrible mother, like I'm letting my babies down. (Nyantakyi-Frimpong 2021: 7)

The impact of climate change on nutrition is, clearly, highly gendered.

The same issues emerge from qualitative research undertaken (using interviews and focus group discussions) in drought-affected communities in south and west Zambia (Rosen *et al.* 2021). Results show very clearly how drought leads not only to food insecurity but also highly gendered effects, with women and young girls becoming vulnerable as reduced household incomes get supplemented by earlier marriage and transactional sex. As one respondent puts it: "People used to get married at the right age, like 25 years and above. But this time, children as young as 12 years old are forced into early marriages in order for their parents to survive." Another reports: "If she goes to an area to ask for food, when she returns, the man now starts asking her, "What if we agree like this?" [trade sex for food]. Because she wants to feed her children back home, she agrees" (Rosen *et al.* 2021: 8). Health service use (such as antenatal health screening) is declining, and childbirth outside clinics increasing, as services become unaffordable and women prioritize securing food over attending to healthcare needs. As Rosen *et al.* (2021: 9) put it, "[S]evere and persistent droughts aggravated underlying socio-economic and structural vulnerabilities for women and girls."

In concluding this section, we should note that Swinburn *et al.* (2019) consider climate change to be a pandemic because of its major effects on human and planetary health. But they link it with both obesity and undernutrition (also described as pandemics) to propose a global syndemic – the synergy of pandemics discussed in Chapter 8. The biggest threat to human health from climate change is, they assert, undernutrition. Papworth, Maslin and Randalls (2015) question whether, as some have argued, climate change is the biggest threat to global health. For these authors, threats from non-communicable diseases will figure much more prominently over the next 30 years.

Climate change and impacts on the mental health of Indigenous peoples

Because of climate change, Indigenous peoples may witness both adverse physical and spiritual losses to their land, impacting on health and wellbeing. Although it draws on a vast array of published scientific research, the IPCC report also recognizes the value of Indigenous knowledge in helping to reduce risks from human-induced climate change (IPCC 2022: 7, ch. 7).

As noted by Bryson et al. (2021), climate change affects the food security of many Indigenous peoples, whether the food is wild fruit consumed by Inuit and First Nations in Canada, or hunted game for groups in South America. The authors collaborated with rural communities (including Indigenous Batwa) in Uganda to explore the risks of climate change (Figure 9.3). Most of the participants acknowledged that long-term changes in climate were affecting food security. Heat waves and increasingly intense precipitation are "heightening health risks for pregnant women who must work outside for food and making it easier to enter and harder to exit the cycle of ill health and poor food security" (Bryson et al. 2021: 11).

It would be wrong to paint a picture of Indigenous peoples as uniformly helpless in the face of climate change. Indeed, adopting some traditional technologies can potentially help mitigate some of the effects, as shown by research undertaken in Ecuador by Carrasco-Torrontegui et al. (2020), who claim that Indigenous contributions to addressing climate change and food security are poorly recognized. Such sustainable technologies include terraces to prevent soil erosion and water loss, raised beds whose water channels are used for aquaculture, and reservoirs for water storage. All of these can help provide valuable nutrient-rich foods while simultaneously preserving the cultural heritage.

The pathways leading to poor mental health as a result of climate change include exposure to severe weather events, disrupted social networks and loss of place attachment. As Middleton et al. (2020) note, the impacts are unequally distributed from place to place and among social groups. They include populations that rely on the environment for farming, hunting and fishing, as well as those whose resources (physical or economic) limit their ability to adapt. Their study of Inuit in Labrador, Canada, makes this very clear. "Given the importance of seasonal cycles in weather across the region, almost all participants reported climatic observations related to seasonality, noting 'shorter' winters, 'spring thaws coming faster,' and 'in the fall it seems like it takes so long to freeze

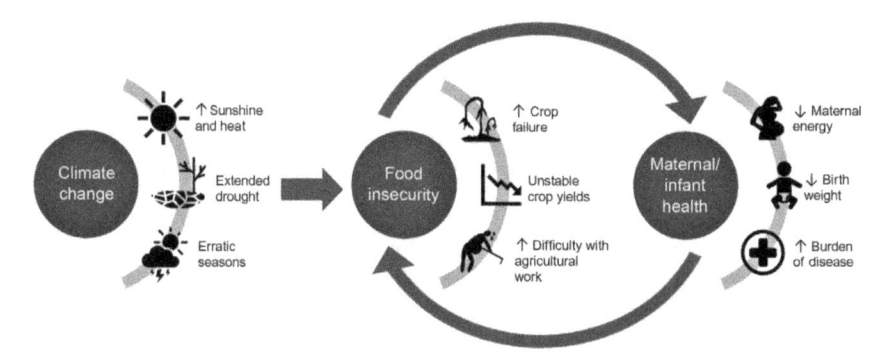

Figure 9.3 Pathways of climate change, food security and maternal/infant health reported by mothers in rural Uganda
Source: Bryson et al. (2021).

up now, it seems like the fall drags on forever'" (Middleton *et al.* 2020). Although many of the comments by Inuit people related to weather, the authors claim that these anticipate likely mental health consequences of climate change. Other research by this group (Cunsolo *et al.* 2020) draws on their concept of "ecological grief", the "grief felt in relation to experienced or anticipated ecological losses, including the loss of species, ecosystems and meaningful landscapes due to acute or chronic environmental change" (Cunsolo *et al.* 2020: 32). Quantitative research (Middleton *et al.* 2021) confirms that, for Inuit in Labrador, there are statistically significant associations between high temperatures (that is, above −5°C) and an increased rate of visits to mental health clinics, although these associations vary from place to place. The stresses faced by these communities are overlain on long-standing inequities and intergenerational trauma (see above, pp. 52–5).

Climate change and population displacement

Although the consequences of climate change for food security are well known, it is difficult to disentangle the potential impact of climate change on population movement, since this is also impacted by political conflict and violence and the economic aspirations of vulnerable people in the Global South. Although Chapter 5 looked in detail at people on the move, what can we say with confidence about the particular impact of climate change on population displacement? As the example in Box 9.2 suggests, caution must be exercised, and some (such as Benveniste, Oppenheimer & Fleurbaey 2022) argue that, rather than it enabling displacement, it may inhibit it; vulnerable people in the poorest countries may find that they are trapped by resource constraints, placing them in yet further danger of severe deprivation.

BOX 9.2 CLIMATE CHANGE, POPULATION DISPLACEMENT AND POLITICAL CONFLICT: A CAUTIONARY TALE

According to the UNHCR, almost 14 million Syrians are thought to have been displaced by the civil war that began in 2011; half are refugees and half have been internally displaced (USA for UNHCR 2023). In their overview of climate change, migration and civil conflict, Balsari, Dresser and Leaning (2020: 407) assert that "the preponderance of the literature suggests that climate change has contributed to civil strife in Syria". Such a claim may be exaggerated.

Abel *et al.* (2019) note that drought and water scarcity, along with poor management of water resources, have led to crop failure, and hence the mass movement of rural population to Syrian cities, leading to urban overcrowding, unemployment and political unrest. Their own empirical research supports a model that suggests that drought and consequential crop failure lead to political unrest and resource conflict, and thence to migration (Figure 9.4), although only for a limited time period.

Selby *et al.* (2017) look closely at the evidence of an association between anthropogenic climate change, drought, population displacement and the subsequent civil war. They conclude that there is no such evidence: the drought caused much less migration than previously alleged, nor did such pressures contribute to the onset of civil war. Although drought did affect northeast Syria between 2006 and 2009, other parts of the country were relatively unaffected. Government attempts to liberalize the economy (resulting in poorer support for agriculture) were key factors in forcing people to move, and large-scale population movement was occurring well before the drought. For these authors, the links between climate change, population movement and conflict must be treated with caution.

Despite the need for caution, there are plenty of human stories suggesting a connection between global warming, food insecurity and consequential migration. Somalia provides a telling example, with episodes of drought occurring regularly during the last 30 years. Yuen, Warsame and Checchi (2022) show that internal population displacement has had little to do with armed conflict and more to do with very low rainfall. There were clear lag effects: very low rainfall was followed three months later by high rates of out-migration

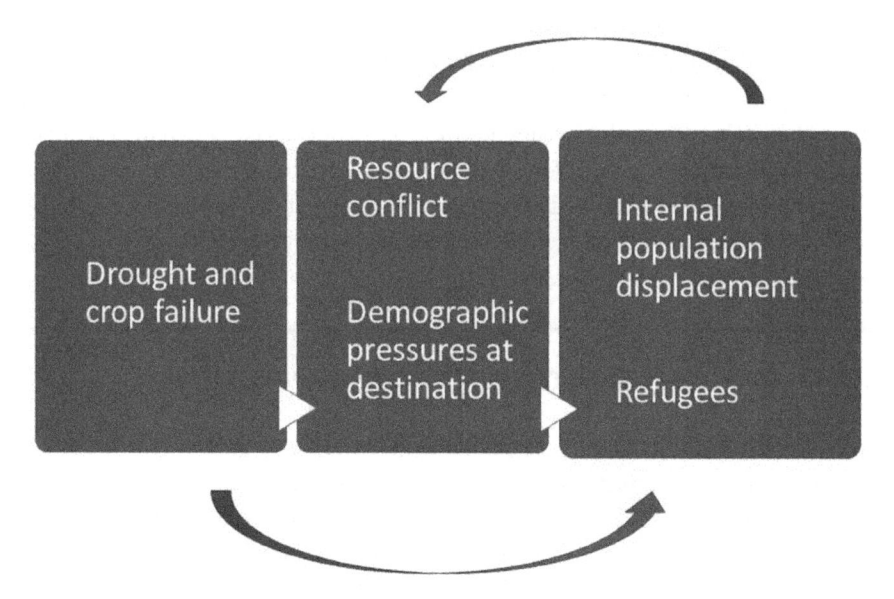

Figure 9.4 Conceptual model of climate, conflict and migration
Source: Author's own, based on Abel *et al.* (2019).

from drought-affected districts. Periods of severe drought disrupted harvests and affected livestock health, leading to farming families moving to prevent food insecurity.

We need to hear from those directly affected. A report in the British newspaper *The Guardian* in April 2022 quotes Abdirahman Nur Hassan, a member of a local drought committee in Somalia, who asserts that attempts to reach Europe are increasingly common and that drought is a primary factor: "If this drought continues, things will get worse, the remaining animals will die, and the majority of people living in this area will end up displaced" (Hayden 2022). The Norwegian Refugee Council reports that 745,000 Somali people were displaced by drought in the first quarter of 2022. Across the globe, the IPCC suggests that 20 million people have been displaced by drought each year since 2008 (Hayden 2022).

A critical examination of possible links between climate change and population displacement is also needed when looking at the possible impacts of sea level rise on small islands. A good example is research by Kelman *et al.* (2019) on the views of people living in the low-lying Maldives, in the Indian Ocean. As the authors assert, there is a danger of the literature being "sometimes subsumed by a discourse of drowning/disappearing islands leading to the islanders desperately preparing to become 'climate change refugees'" (Kelman *et al.* 2019: 293). Interviews of 113 Maldivians offered a range of responses, but most had more concerns about contemporary issues such as income, employment and education rather than potential future impacts of climate change and an urge to relocate. Similarly, Farbotko (2019: 256) refers to some communities in low-lying Pacific islands as "voluntarily immobile", with members choosing to stay in areas that are becoming technically uninhabitable, often for cultural, emotional and spiritual reasons, thereby enabling their ontological security. Although this may not seem rational in the context of the impending physical threats of sea level rise, it can be seen as a form of resistance to those voices in the Global North that are louder than those of Indigenous Pacific islanders.

Issues of relocation, whether forced or voluntary, relate to much broader issues of adaptation to climate change. Although this is a pressing issue, it is important to take a critical stance when considering climate change adaptation (Box 9.3).

BOX 9.3 ADAPTING TO CLIMATE CHANGE: A CRITICAL PERSPECTIVE

Strategies to *mitigate* the effects of climate change, most obviously reducing greenhouse gas emissions, are beyond the scope of this book, as are strategies to *adapt* to climate change – changing behaviours, technologies, infrastructure, warning systems, and so on. For attempts to gather evidence on climate change adaptation, the Global Adaptation Mapping Initiative (https://globaladaptation.github.io) may be consulted, as should the series of *Adaptation Gap* reports by the United Nations Environment Programme (UNEP) (latest UNEP 2021). However, it is worth casting a brief and critical eye on climate change adaptation.

Taylor (2014: 3) has noted that "Adaptation now!" has become a "shared refrain of international institutions, national governments, non-governmental organisations and scholars", a refrain that serves to organize and legitimize practical interventions. His preferred political ecological approach abandons a binary separation of climate and society and sees climate change adaptation as linked to issues of power and security. For example, in a study of drought in the semi-arid Deccan plateau in southern India, he examines the impacts of climatic change on marginal and smallholder farmers struggling to adapt and survive in the wake of very unequal distributions of risk and security. Echoing Singer's introductory quote, he suggests that "we need to explicitly foreground ways to collectively deleverage a global capitalist order that is predicated upon the unending accumulation of productive forces and consumptive practices that give rise to the deadly metabolisms inherent to climatic change" (Taylor 2014: 8).

Concluding remarks

After Covid-19, Romanello *et al.* (2021) paint a bleak future picture of climate change for population health. They suggest that recovering from the pandemic is generating new increases in greenhouse gas emissions, returning these to pre-pandemic levels. Although the impacts of climate change have been known for many years, efforts to adapt and build resilience have been piecemeal and unequal, with the poorest countries least able to respond to the challenges.

Understanding the impacts of climate change demands an engagement with many of the issues and concepts used in this book. Among these are concerns with environmental justice and inequity. As many writers have asserted, there are huge inequities in terms of those countries contributing to climate change and those on the receiving end of its impacts. Island nations contribute tiny proportions of global greenhouse gas emissions, but are among the most vulnerable (Tukuitonga & Vivili 2021). There is a lack of "climate justice".

The likely future burdens of climate change represent a form of structural and slow violence visited on those countries and communities least able to manage the threats. As with Covid-19, any suggestion that "we are all in this together" is absurd. Equally, climate change is the prime example of what Ulrich Beck meant when he spoke about a world risk society, and what others understand by globalization. Globalization has meant that production and consumption, particularly in the Global North, have seen their effects unfold in places far away.

In an earlier chapter I set out the Sustainable Development Goals. But these contain internal contradictions. For example, although SDG 13 is a call to keep global temperature change to under 2°C, and advocates "sustainable patterns of production and consumption", SDG 8 is focused on economic growth and calls for annual GDP growth of at least 7 per cent in low-income countries. Both, regrettably, are flights of fancy.

Further reading

Chapter 7 of the IPCC report (IPCC 2022) is very comprehensive and contains numerous references. See Box 7.1 in that chapter for a review of climate change and Indigenous health. Along with Sultana (2022), see Tilley *et al.* (2022) on anti-racist ecologies and climate change.

The lengthy article by Swinburn *et al.* (2019) looks at the syndemic of climate change, undernutrition and obesity, and is a valuable contribution. Although slightly dated, Curtis and Oven (2012) provide a useful overview of links between climate change and health geographies, while Papworth, Maslin and Randalls (2015) question whether climate change policy should be prioritized over addressing poverty and inequality. Helldén *et al.* (2021) have an excellent recent overview of links between climate change and child health. Rupasinghe, Chomel and Martínez-López (2022) have an up-to-date overview of the likely consequences of climate change for a wide range of insect-borne and water-borne infections.

Most useful, perhaps, are the regular and very thorough *Lancet Countdown* reviews of health and climate change; see, for example, Romanello *et al.* (2021) and, more recently, the 2022 report. Consulting articles regularly in the journal *Nature Climate Change* is also recommended.

Last, readers should be aware of research suggesting that there is clear evidence that various climate tipping points are likely to be reached, or have already been reached. These include the collapse of Greenland's ice cap and permafrost thawing. See McKay *et al.* (2022). All will have major impacts on human health.

10
CONCLUSIONS: GLOBAL HEALTH AND CROSS-CUTTING THEMES

> I argue that the plagues of the twenty-first century are not an isolated set of diseases ... but an ever more complex array of interacting diseases, the spread of which is being driven by the dual (and themselves interacting) forces of globalism and global warming, both of which are shaped by human action, decision making, and the unequal and often oppressive structure of social relations.
>
> Singer (2010: 25)

In their introduction to health geographies, Brown *et al.* (2018) refer to "five cross-cutting critical themes" that thread throughout their chapters. In a book on global health it is inevitable that my own choice of cross-cutting themes overlaps with theirs, but the material I have covered leads me to suggest some alternatives (Table 10.1).

In this concluding chapter I want to use these themes to summarize what has been presented in earlier chapters.

Governance/security

Power, whether political or economic, is held by a wide range of actors – individual, organizational, state or international – with complex relations between different levels. Actors can include NGOs, government departments, private companies, landowners,

Table 10.1 Cross-cutting themes in health geography and global health

Health Geographies: A Critical Introduction[1]	Global Health: Geographical Connections
Neoliberalism	Governance/security
Inequality	Inequality/injustice
Biopolitics	Political ecology/syndemics
Resistance/resilience/care	Structural violence
Globalization/urbanization	Globalization/networks

Note: [1] From Brown *et al.* (2018: ch. 1).

and so on. All these, often networked together, dictate access to resources, including healthcare, that in turn shape health or its absence. The power contained within these networks may contribute to structural violence, as noted later. Yet states and international organizations alike can, as Farmer (2005: 142) has pointed out, be relatively power*less*.

> The impotence of international bodies in the face of generally recognised problems, their inability to effect solutions, stems from the self-interest of those who stand to benefit from their oppression of other human beings. In each major problem there is broad recognition of both the moral intolerableness and the political non-viability of the existing situation, coupled with a lack of capacity to respond. (Farmer 2005: 142)

We saw in Chapter 3 that there is a debate among political geographers and international relations researchers as to the relative power exercised by the state and international bodies such as the WHO, notably in addressing infectious disease spread. The Global Health Security Index remains resolutely fixated on security from infectious disease. But countries in the Global North have their own, internal, issues of security. A broad definition reveals that countries such as the United Kingdom have serious challenges in securing their poor from the costs of energy and food shortages, and the wider population from environmental mismanagement; poor water quality and sanitation are issues not only in the Global South. Taking a much narrower view of security, we have seen how borders can be both porous and policed. The porosity allows infections to spread from place to place, disrespecting political boundaries. The policing seeks to control movement, either to prevent disease spread or, more broadly, to stop people moving from place to place to secure for themselves a better (and healthier) life.

As noted in Chapter 3, health security has often been seen as securing countries in the Global North from threats (usually from viruses) originating in the Global South. Climate change reverses this dynamic, since, although there are threats (such as from sea level rise) to countries in the Global North, it is demonstrably the case that low-lying island countries in the Pacific are at risk from what the North has been doing for many years in adding to global warming via emissions of CO_2. The devastating floods in Pakistan in 2022 are widely thought to have been caused not only by heavy monsoon rainfall but also by glaciers in northern Pakistan melting. The consequences of such flooding, affecting at least 33 million in the provinces of Sindh and Balochistan, are major outbreaks of infections such as cholera and other waterborne diseases.

Much has been made of the negative impacts on global health from neoliberal agendas. But it would be a mistake to suggest that all the power is invested in the capitalist economies and major international bodies such as the World Bank. There is plenty of evidence to suggest that global players such as Russia and China, both (quasi-)authoritarian states, have contributed to, and continue to contribute to, global ill health, whether by disrupting food supplies to those in need (a result of Russia's invasion of Ukraine) or a continuing reliance on coal (in China). It is worth repeating the words of Schrecker (2020) from Chapter 3: "Global health politics must be understood with the recognition that suffering can be inflicted and lives destroyed by remote control: by choices made half a world away, by people and within institutions that have no contact with those affected and are in no way accountable to them." Not all such institutions are based in Washington, DC, or Geneva.

Inequality/injustice

We considered briefly in Chapter 2 the Sustainable Development Goals, noting their inattention to both global power relations and the underlying structural causes of poverty and ill health. Moreover, the data needed to track SDG progress are quite limited. Swain (2017) notes that, although data on mortality are assumed to be widely available, two-thirds of the 75 countries that account for more than 95 per cent of maternal, infant and child deaths lack registries of births and deaths. Data on Indigenous groups is invariably missing.

As noted in the Introduction, it is well established that who you are (your age, gender, income and social class), as well as where you live, all shape health and wellbeing. The social and political determinants of health are expressed spatially, whether in small neighbourhoods or wider regions, and this is true in both the Global South and Global North. For example, access to healthcare is challenging in the United Kingdom and United States, whether as a result of staff shortages, lack of insurance cover or unequal geographic coverage. Inequality (or, more accurately, inequity) is rife across the globe, and, as revealed in Chapters 2 and 3 particularly, it shapes health outcomes. There may be a "happy planet" (Box 10.1) for some, but not for many.

BOX 10.1 A HAPPY PLANET?

In 2006 the New Economics Foundation devised a "Happy Planet Index" (HPI), now overseen by the Wellbeing Economy Alliance (https://happyplanetindex.org/about-us). The HPI seeks to integrate national data on life expectancy, subjective wellbeing and ecological footprint, with the aim of "comparing countries' progress towards the goal of providing long-term wellbeing for all, without exceeding the limits of the planet's resources". It seeks to measure how well countries "deliver long, happy lives using limited environmental resources". As the team behind the index acknowledge, average life expectancy masks considerable within-country variation, subjective wellbeing is only imperfectly captured by a ten-point scale, and ecological footprint is based solely on the average amount of land needed to sustain a country's typical consumption patterns. Given these caveats, it is interesting to note that countries in Central and South America dominate the top ten in 2020: Costa Rica, Vanuatu, Colombia, Switzerland, Ecuador, Panama, Jamaica, Guatemala, Honduras and Uruguay. Countries in the Global North, notably in Europe, score highly in terms of life expectancy and wellbeing, but poorly on ecological footprint.

The scale of poverty, absolute deprivation and disease in large parts of the Global South dwarfs that in the Global North. Sadly, the continuing dominance of a neoliberal agenda, championed by powerful interest groups and politicians, will do nothing to ameliorate the appalling conditions faced by the world's poor and marginalized. Such marginalization is represented and evidenced clearly among Indigenous groups in, for example, Australia and North America, where forced removal and land dispossession have led to trauma, which has been transmitted through families and communities to the present day (see above, pp. 53–5). In his powerful testimony of the lives and health of people in Haiti, Farmer (2005: 158) reminds us that "modern-day Haitians are the descendants of a people enslaved in order to provide our ancestors with cheap sugar, coffee, and cotton". Their enslavement has itself been transmitted down the generations to limit the life chances and wellbeing of contemporary Haitians.

Social justice means protecting human rights, permitting access to essential resources, allowing people a voice (participation) and the elimination of gross inequity. Health justice means seeing health as a basic human right, not simply a commodity that can be bought only by those who can afford to pay for services and care. Khosla, Allotey and Gruskin (2020: 1) note that, "with the waning of, or growing ennui from the shock of the pandemic, the world seems ready to slip back into 'avarice' with little thought". For them, global health is a matter of social justice and human rights (Box 10.2).

Beyond concerns with social justice and human rights, some writers have begun to speak of "planetary justice", a concept that embraces climate justice (the recognition that the impacts of climate change are borne unequally). A focus of planetary justice is on securing both environmental quality and human rights and equity, and involves not just international actors but also local communities, particularly Indigenous groups. It "attributes both environmental degradation and poverty to the effects of large-scale industrialism, corporate control of the economy and hegemonic globalization of the ideologies of capitalism and consumption" (Kashwan *et al*. 2020). These authors argue that planetary justice should attend to the needs of the poor and involve them in decision-making. San Martín and Wood (2022) agree, suggesting that planetary justice requires "pluralizing": recognizing and listening to the voices of those who are marginalized and

BOX 10.2 HEALTH AND HUMAN RIGHTS

The WHO constitution states that "[t]he enjoyment of the highest attainable standard of health is one of the fundamental rights of every human being without distinction of race, religion, political belief, economic or social condition" (WHO n.d.d). Much of the present book has been concerned to show how, with an emphasis on the Global South, these health rights – and, crucially, their determinants – are spatially and environmentally, as well as socially, patterned.

These rights are encapsulated in the set of Sustainable Development Goals and their various targets and indicators. In particular, SDG 13 urges action to tackle climate change, which has clear impacts (via food chains, water, sanitation, and so on) on the right to good health. As Patterson (2022) notes, there is a crucial role to be played by individuals and social groups in demanding these rights. However, the spaces within which people can meet to protest are being curtailed, with concerns that "restrictions on civic space will continue after the COVID-19 pandemic, with negative implications for civil society responses to other global challenges, including the climate crisis" (Patterson 2022: 116). In the United Kingdom we are already witnessing government attempts to curb protest (Liberty 2022).

For further reading on human rights and health, see papers in the *Health and Human Rights Journal*.

---

oppressed. As we saw in Chapter 1, the same arguments have been made concerning the decolonizing of global health.

All these "justices" are, of course, interconnected. The deep structural determinants of climate change, ecological disruption, pandemic disease and health inequity are overlapping. Many of these consequences, as we have seen, are revealed most clearly in the Global South. Solomonian and Di Ruggiero (2021: 2) put this in the starkest possible terms:

> Those that are disproportionately affected by climate change (through consequences such as floods, droughts, fires, and conflict) are the same who have been exploited, displaced, marginalized, and murdered to concentrate wealth for the elite, whose actions have further driven the degradation of the environment.

Planetary justice and health justice are closely linked to notions of planetary health and One Health. Planetary health is concerned with the health of human populations and the natural world alike. One Health seeks to explore health and disease as a set of relations between humans, animals and plants, within their social and environmental contexts, although one has to ask how this differs from the disease ecology framework used by geographers for many years (Chapter 1, pp. 2–6). Ferdowsian (2021: 2) criticizes the concept for "being too anthropocentric and for failing to include adequate attention to human and nonhuman rights and well-being". For other writers, such as Davis and Sharp (2020) and Hinchliffe (2017), One Health neglects the "entanglements" of human and animal populations – the assemblages that were referred to in Chapter 1.

Climate change has impacts that are felt acutely by women in the Global South, leading some ecofeminist writers, such as Terry (2009), to argue that "there is no climate justice without gender justice". An ecofeminist approach also argues that planetary health demands attention to the interconnectedness of the human and non-human worlds and requires an ethics of care rather than destruction of complex relations in the search for profit. The breaching of "planetary boundaries" (Steffen *et al.* 2015), whether these relate to climate, land degradation, freshwaters or others, will have consequences for human health that are already apparent, as shown in earlier chapters. Writing on behalf of the United Nations Office for Disaster Risk Reduction, Cernev (2022) has recently suggested that such breaching may lead to total societal collapse.

Political ecology/syndemics

Many of the studies considered earlier have drawn on a political ecology framework, one that brings together both political economy and ecology to form a critical social science. Faye and Braun (2022) link human health to the health of soils, showing how the sustained monocropping of peanuts in Senegal, west Africa, removes nutrients and encourages the proliferation of aflatoxins (fungal infections), which enter the food chain, affecting both food security and possible disease. Importantly, substantial Chinese investment in Senegal has led Chinese companies exporting to China and Europe, while climate change increases the risk of aflatoxins in the crop. For these authors, a political ecology approach places disease threat both in an historical context of the colonial cash crop production of peanuts and in a contemporary context of market liberalization, soil degradation and climate change.

The notion of a syndemic, introduced in Chapter 8, proves useful in illuminating ill health and disease. The social context or environment within which people live operates in conjunction with individual-level factors to provide a backcloth on which coexisting diseases are mapped. The suffering that results is borne unequally by marginalized communities, who are subject to the structural violence brought about by neoliberalism and its antecedents. Herring and Swedlund (2010) put it well: "The syndemic approach requires that researchers look beyond individual infections to consider how they may be *capacitated by the presence of other diseases and conditions and sustained by social inequity* and the unjust exercise of power, which channels and sustains damaging disease clusters in disadvantaged populations" (Herring & Swedlund 2010: 5, emphasis added).

Singer, who first originated the term "syndemic", has broadened the concept to include environmental change, including global warming; hence he speaks of ecosyndemics. As the introductory quote suggests, he uses this new term to link globalization, climate change, disease, power and inequality into a broad ecology of global health. Tallman *et al.* (2022) agree that ecosyndemics refer to disease interactions arising from the environmental changes brought about by human beings. In their important research on dam and highway construction (in Brazil and Peru respectively), Tallman and her colleagues suggest that such infrastructure sets the conditions for vector-borne infections, commercial exchanges for sex and psychosocial stress, threatening immune systems and leading to increased population morbidity.

Structural violence

I suggest that another critical concept for studying global health is structural violence: violence on the weak and powerless, undertaken by those in power (Box 1.1). It may be unintended, but it is nonetheless real. The concept applies in many settings, such as in relation to the harm done to women lacking food, maternal care in resource-poor environments or the racial discrimination suffered by people from Africa and the Middle East, desperate for safety in European countries. It also has purchase in understanding the impacts of climate change.

Bonds (2016) notes that, although some researchers on climate change argue that it correlates with increased levels of physical violence, others contest this. For example, Barnett (2019) refers to "climate-resilient peace": peace persists during climate change. Bonds prefers to argue that climate change is itself a form of violence – structural violence – perpetuated by powerful interest groups who benefit from, and sustain, a dependence on fossil fuel (Bonds 2016). Such *structural* violence can itself be overcome only by *structural* changes to production and consumption. Nicoson (2021) too draws on the concept of climate resilient peace and links climate change not to physical violence but to structural violence, arguing for "degrowth", which reduces production and consumption and focuses more on the redistribution needed to reduce structural violence and enhance global equity. She argues that the power structures that produce and reproduce such inequity need disrupting.

Some years ago Young (1990) referred in her work on justice to "five faces" of oppression, which she listed as exploitation, marginalization, powerlessness, cultural imperialism and violence (by which she meant physical or sexual violence). Clearly, the first three of these can be bracketed as structural violence. People who are oppressed because of colour, gender or political beliefs, for example, suffer consequences for their health, well-being and risk of premature mortality, all of which vary from place to place.

Globalization/networks

Earlier chapters considered social and political determinants – the upstream factors that shape health. Labonté and Schrecker (2007) refer to globalization as the "quintessential upstream variable". Globalization enables these determinants, such as by major corporations engaging in the widespread marketing of tobacco and foods that contribute to non-communicable diseases. It is perpetuated through the influence of organizations such as the World Bank and International Monetary Fund, which argue that the integration of countries in the Global South into the global economy, via international trade, can contribute to poverty reduction (and thereby improved population health). But, as we saw in Chapter 3, the structural adjustment programmes imposed on some countries have hardly witnessed such improvements, with evidence suggesting reductions in health service expenditure and provision. Such reductions invariably have unequal impacts that are spatially differentiated and felt locally.

The global and local are therefore interdependent. What happens locally, in particular places, is invariably affected by global economic and political processes. Equally, what happens in such places can have wide-ranging effects on social life and human health. The obvious contemporary example is the (likely) emergence of SARS-CoV-2 in the seafood market of Wuhan and its global, devastating spread.

Consequently, the clearest illustrations of the geographical connections alluded to in this book's subtitle come from mapping and analysing the flows of material, people and infections from place to place. These have been covered in detail in Chapters 5, 6 and 8. We can model these flows and perhaps use such models to predict future movements, notably possible disease spread. The Covid-19 pandemic has re-energized such modelling. Valuable as these models sometimes are, I argue that we need to dig deeper to uncover the structure of these networks and the causes of such flows. To do so requires an historical, as well as a geographical, imagination. For example, as Paul Farmer's last book – on Ebola – asserted, "West African epidemics and social responses to them can't be fully comprehended without knowledge of the region's long entanglement with Europe and the Americas" (Farmer 2020: xxvi). As he demonstrates convincingly, one sort of flow generates others; the extraction and movement of minerals (diamonds, titanium, bauxite) and the movement of war refugees across borders all contributed to the spread of Ebola virus in upper west Africa. In his own words:

> Armed conflict and forced labor are among the causes of the causes that rolled out the red carpet for rapid human-to-human spread of Ebola in Upper West Africa. These wars weren't fought, nor labor forced, for strictly local gain. An understanding of today's West African epidemics, and social responses to them, required consideration of the region's integration into an economic web spanning the Atlantic Ocean. (Farmer 2020: 186)

However, although some networks help spread disease and others extract resources to benefit those in parts of the Global North, other networks – of researchers and activists – are needed to address global health issues. Research evidence is of course needed to inform policy. But in many instances we have more than enough evidence. We do not need yet more empirical research to reveal stark inequalities or to illustrate that poor health is caused by malnutrition, lack of sanitation or clean drinking water. *We have the evidence*, as revealed by, literally, thousands of research papers and dozens of books. What is needed is the political will to address the structural violence that affects millions worldwide. What is required is a coalition of the willing: a coalition that brings together health practitioners, academics from many disciplines (including geography) and activists – to prioritize the health needs of the oppressed and disadvantaged. We need to acknowledge that inequality is both a cause and a consequence of much ill health and disease, and this is as true of the Global North as of the Global South.

In tribute to the late Paul Farmer, I close with a quote from a chapter he co-authored: "Global health equity is a noble ambition, but it remains only a beginning to the pursuit of a more just, fair society that allows our children, wherever they are born, a decent shot at a decent life" (Basilico *et al.* 2013b: 353).

REFERENCES

Abel, G. *et al.* 2019. "Climate, conflict, and forced migration". *Global Environmental Change* 54: 239–49.
Abimbola, S. & M. Pai 2020. "Will global health survive its decolonisation?". *The Lancet* 396: 1627–8.
Abu, T., E. Bisung & S. Elliott 2019. "What if your husband doesn't feel the pressure? An exploration of women's involvement in WaSH decision making in Nyanchwa, Kenya". *International Journal of Environmental Research and Public Health* 16 (10). DOI: 10.3390/ijerph16101763.
Adams, V. (ed.) 2016. *Metrics: What Counts in Global Health*. Durham, NC: Duke University Press.
Adey, P. *et al.* (eds) 2015. *The Routledge Handbook of Mobilities*. London: Routledge.
Adjaye-Gbewonyo, K. *et al.* 2016. "Income inequality and depressive symptoms in South Africa: a longitudinal analysis of the National Income Dynamics study". *Health & Place* 42: 37–46.
Affun-Adegbulu, C. & O. Adegbulu 2020. "Decolonising global (public) health: from Western universalism to global pluriversalities". *BMJ Global Health* 5 (8). DOI: 10.1136/bmjgh-2020-002947.
AgênciaBrasil 2021. "BNDES president says Brazil faces challenge of fighting inequality". 28 September. https://agenciabrasil.ebc.com.br/en/economia/noticia/2021-09/bndes-president-says-brazil-faces-challenge-fighting-inequality.
Ahmad, S., M. Khan & M. Haque 2018. "Arsenic contamination in groundwater in Bangladesh: implications and challenges for healthcare policy". *Risk Management and Healthcare Policy* 11: 251–61.
Ahmed, K. 2022. "'Like an open prison': a million Rohingya refugees still in Bangladesh camps five years after crisis". *The Guardian*, 23 August. www.theguardian.com/global-development/2022/aug/23/five-years-rohingya-refugees-2017-bangladesh-myanmar-military-crackdown.
Aked, J. *et al.* 2008. "Five ways to wellbeing". New Economics Foundation, 22 October. https://neweconomics.org/2008/10/five-ways-to-wellbeing.
Akese, G. & P. Little 2018. "Electronic waste and the environmental justice challenge in Agbogbloshie". *Environmental Justice* 11 (2): 77–83.
Akese, G., U. Beisel & M. Chasant 2022. "Agbogbloshie: a year after the violent demolition". African Arguments, 21 July. https://africanarguments.org/2022/07/agbogbloshie-a-year-after-the-violent-demolition.
Akinwotu, E. 2022. "Sooty hands and damaged lungs: the toll of Nigeria's illegal refineries". *The Guardian*, 29 May. www.theguardian.com/environment/2022/may/29/sooty-hands-and-damaged-lungs-the-toll-of-nigerias-refineries.
Aldred, T.-L. *et al.* 2021. "Mining sick: creatively unsettling normative narratives about industry, environment, extraction, and the health geographies of rural, remote, northern, and Indigenous communities in British Columbia". *The Canadian Geographer* 65 (1): 82–96.

REFERENCES

Alhaffar, M. & S. Janos 2021. "Public health consequences after ten years of the Syrian crisis: a literature review". *Globalization and Health* 17 (1). DOI: 10.1186/s12992-021-00762-9.

Ali, H. & R. Keil (eds) 2008. *Networked Disease: Emerging Infections in the Global City*. Chichester: Wiley-Blackwell.

Allen, L. *et al.* 2022. "Conflict as a macrodeterminant of non-communicable diseases: the experience of Libya". *BMJ Global Health* 7 (Supp. 8). DOI: 10.1136/bmjgh-2021-007549.

Allin, P. & D. Hand 2014. *The Wellbeing of Nations: Meaning, Motive and Measurement*. Chichester: John Wiley.

Allison, E. *et al.* 2009. "Vulnerability of national economies to the impacts of climate change on fisheries". *Fish and Fisheries* 10 (2): 173–96.

Allport, T. *et al.* 2019. "'Like a life in a cage': understanding child play and social interaction in Somali refugee families in the UK". *Health & Place* 56: 191–201.

Amin, A. & L. Richaud 2020. "Stress and the ecology of urban experience: migrant mental lives in central Shanghai". *Transactions of the Institute of British Geographers* 45 (3): 862–76.

Amodu, O., M. Richter & B. Salami 2020. "A scoping review of the health of conflict-induced internally displaced women in Africa". *International Journal of Environmental Research and Public Health* 17 (4). DOI: 10.3390/ijerph17041280.

Amrith, S. 2009. "Health in India since independence", Working Paper 79. Manchester: Brooks World Poverty Institute.

Amzat, J. & O. Razum 2022. *Globalization, Health and the Global South: A Critical Approach*. Abingdon: Routledge.

Anderson, B. *et al.* 2012. "On assemblages and geography". *Dialogues in Human Geography* 2 (2): 171–89.

Anderson, W. 2021. "The model crisis, or how to have critical promiscuity in the time of COVID-19". *Social Studies of Science* 51 (2): 167–88.

Andrews, G. 2018. *Non-Representational Theory and Health: The Health in Life in Space-Time Revealing*. Abingdon: Routledge.

Andrews, G. 2021. "Re-imagining world: from human health in the world to 'all-world health'". *Health & Place* 71. DOI: 10.1016/j.healthplace.2021.102620.

Andrews, G. *et al.* (eds) 2021. *COVID-19 and Similar Futures: Pandemic Geographies*. New York: Springer.

Anjara, S. *et al.* 2017. "Stress, health and quality of life of female migrant domestic workers in Singapore: a cross-sectional study". *BMC Women's Health* 17 (1). DOI: 10.1186/s12905-017-0442-7.

Arcaya, M., A. Arcaya & S. Subramanian 2015. "Inequalities in health: definitions, concepts, and theories". *Global Health Action* 8. DOI: 10.3402/gha.v8.27106.

Atkinson, S. 2016. "Health and wellbeing". In *International Encyclopaedia of Geography: People, The Earth, Environment and Technology*, D. Richardson *et al.* (eds), 15 vols. Chichester: Wiley-Blackwell. DOI: 10.1002/9781118786352.wbieg0770.

Baldé, C. *et al.* 2022. *Global Transboundary E-Waste Flows Monitor: 2022*. Bonn: UNITAR.

Balsari, S., C. Dresser & J. Leaning 2020. "Climate change, migration, and civil strife". *Current Environmental Health Reports* 7 (4): 404–14.

Bambra, C. 2016. *Health Divides: Where You Live Can Kill You*. Bristol: Policy Press.

Bambra, C., K. Smith & J. Pearce 2019. "Scaling up: the politics of health and place". *Social Science & Medicine* 232: 36–42.

BAN 2019. "Watchdog group uses GPS trackers to discover illegal electronic waste exports from Europe to Africa and Asia". 6 February. www.ban.org/news-new/2019/2/6/gps-trackers-discover-illegal-e-waste-exports-to-africa-and-asia.

Bardosh, K. (ed.) 2020. *Locating Zika: Social Change and Governance in an Age of Mosquito Pandemics*. Abingdon: Routledge.

Bardosh, K. *et al.* 2017. "Addressing vulnerability, building resilience: community-based adaptation to vector-borne diseases in the context of global change". *Infectious Diseases of Poverty* 6 (1). DOI: 10.1186/s40249-017-0375-2.

Barnett, J. 2019. "Global environmental change I: climate resilient peace?". *Progress in Human Geography* 43 (5): 927–36.

Barnett, R. & A. Copeland 2010. "Providing health care". In *A Companion to Health and Medical Geography*, T. Brown, S. McLafferty & G. Moon (eds), 497–520. Oxford: Blackwell.

Bashford, A. (ed.) 2006. *Medicine at the Border: Disease, Globalization and Security, 1850 to the Present*. Basingstoke: Palgrave Macmillan.

Basilico, M. *et al.* 2013a. "Health for all? Competing theories and geopolitics". In *Reimagining Global Health: An Introduction*, P. Farmer *et al.* (eds), 74–110. Berkeley, CA: University of California Press.

Basilico, M. *et al.* 2013b. "A movement for global health equity? A closing reflection". In *Reimagining Global Health: An Introduction*, P. Farmer *et al.* (eds), 340–53. Berkeley, CA: University of California Press.

Battersby, J. 2017. "Eat your greens, buy some chips: contesting articulations of food and food security in children's lives". In *Global Health and Geographical Imaginaries*, C. Herrick & D. Reubi (eds), 195–215. Abingdon: Routledge.

Bayar, M. & M. Aral 2019. "An analysis of large-scale forced migration in Africa". *International Journal of Environmental Research and Public Health* 16 (21). DOI: 10.3390/ijerph16214210.

Bayona-Valderrama, A., T. Acevedo-Guerrero & C. Artur 2020. "Cities with mosquitoes: a political ecology of *Aedes aegypti*'s habitats". *Water Alternatives* 14 (1): 186–203.

Bayram, A. 2022. "We cannot turn away from the ongoing suffering in Syria". NRC, 15 March. www.nrc.no/perspectives/2022/syria-we-cannot-turn-away-from-the-ongoing-suffering-in-syria.

BBC News 2021a "Brazil health service in 'worst crisis in its history'". 17 March. www.bbc.co.uk/news/world-latin-america-56424611.

BBC News 2021b. "Who are the Uyghurs and why is China being accused of genocide?". 24 May. www.bbc.co.uk/news/world-asia-china-22278037.

BBC News 2021c. "COP26: document leak reveals nations lobbying to change key climate report". 21 October. www.bbc.co.uk/news/science-environment-58982445.

BBC News 2021d. "Trafficked to Europe for sex: a survivor's escape story". 23 October. www.bbc.co.uk/news/stories-58994212.

BBC News 2021e. "Foreign aid: who will be hit by the UK government cuts?". 8 November. www.bbc.co.uk/news/57362816.

BBC News 2022a. "Zika virus may be one step away from explosive outbreak". 12 April. www.bbc.co.uk/news/health-61078867.

BBC News 2022b. "Nigeria in trauma after Imo refinery deaths – President Buhari". 24 April. www.bbc.co.uk/news/world-africa-61207441.

Beck, U. 1998. *World Risk Society*. Cambridge: Polity Press.

Beck, U. & C. Lau 2005. "Second modernity as a research agenda: theoretical and empirical explorations in the 'meta-change' of modern society". *British Journal of Sociology* 56 (4): 525–57.

Beisel, U. 2017. "Resistant bodies, malaria and the question of immunity". In *Global Health and Geographical Imaginaries*, C. Herrick & D. Reubi (eds), 114–34. Abingdon: Routledge.

Bell, J. & J. Nuzzo 2021. *Global Health Security Index: Advancing Collective Action and Accountability amid Global Crisis*. Washington, DC: Nuclear Threat Initiative.

REFERENCES

Bell, M. 1995. *Geography and Imperialism, 1820–1940*. Manchester: Manchester University Press.

Bell, S. et al. 2018. "From therapeutic landscapes to healthy spaces, places and practices: a scoping review". *Social Science & Medicine* 196: 123–30.

Bentley, A. et al. 2015. "Malnutrition and infant and young child feeding in informal settlements in Mumbai, India: findings from a census". *Journal of Food Science & Nutrition* 3 (3): 257–71.

Benveniste, H., M. Oppenheimer & M. Fleurbaey 2022. "Climate change increases resource-constrained international immobility". *Nature Climate Change* 12 (7): 634–41.

Benvenisti, E. 2020. "The WHO – destined to fail? Political cooperation and the COVID-19 pandemic". *American Journal of International Law* 114 (4): 588–97.

Berry, I. & L. Berrang-Ford 2016. "Leishmaniasis, conflict and political terror: a spatio-temporal analysis of global incidence". *Social Science & Medicine* 167: 140–9.

Bervell, B. & H. al-Samarraie 2019. "A comparative review of mobile health and electronic health utilization in sub-Saharan African countries". *Social Science & Medicine* 232: 1–16.

Bettampadi, D. et al. 2021. "Vaccination inequality in India, 2002–2013". *American Journal of Preventive Medicine* 60 (Supp. 1): 65–76.

Bharti, N. et al. 2016. "Measuring populations to improve vaccination coverage". *Scientific Reports* 6. DOI: 10.1038/srep34541.

Biehl, J. & A. Petryna (eds) 2013. *When People Come First: Critical Studies in Global Health*. Princeton, NJ: Princeton University Press.

Billing, L. 2021. "'Everything living is dying': environmental ruin in modern Iraq". Undark, 22 December. https://undark.org/2021/12/22/ecocide-iraq.

Birn, A.-E. 2014. "Philanthrocapitalism, past and present: the Rockefeller Foundation, the Gates Foundation, and the setting(s) of the international global health". *Hypothesis* 12 (1). DOI: 10.5779/hypothesis v12i1.229.

Bisung, E. et al. 2015. "Dreaming of toilets: using photovoice to explore knowledge, attitudes and practices around water–health linkages in rural Kenya". *Health & Place* 31: 208–15.

Bloche, M. & E. Jungman 2006. "Health policy and the World Trade Organization". In *Globalization and Health*, I. Kawachi & S. Wamala (eds), 250–67. Oxford: Oxford University Press.

BMA n.d. "People's stories". www.bhopal.org/continuing-disaster/the-bhopal-gas-disaster/peoples-stories.

BMJ 2022. "Pfizer pledges to provide 23 drugs at not-for-profit prices to lower income countries". *British Medical Journal* 377. DOI: 10.1136/bmj.o1329.

Bogard, W. 1989. *The Bhopal Tragedy: Language, Logic, and Politics in the Production of a Hazard*. Boulder, CO: Westview Press.

Bonds, E. 2016. "Upending climate violence research: fossil fuel corporations and the structural violence of climate change". *Human Ecology Review* 22 (2): 3–23.

Bouhenia, M. et al. 2017. "Quantitative survey on health and violence endured by refugees during their journey and in Calais, France". *International Health* 9 (6): 335–42.

Boyd, M., N. Wilson & C. Nelson 2020. "Validation analysis of Global Health Security Index (GHSI) scores 2019". *BMJ Global Health* 5 (10). DOI: 10.1136/bmjgh-2020-003276.

Boyle, M. 2021. *Human Geography: An Essential Introduction*. Chichester: John Wiley.

Breman, A. & C. Shelton 2006. "Structural adjustment programs and health". In *Globalization and Health*, I. Kawachi & S. Wamala (eds), 219–33. Oxford: Oxford University Press.

Brewis, A. et al. 2021. "Household water insecurity and psychological distress in eastern Ethiopia: unfairness and water sharing as undertheorized factors". *Social Science & Medicine: Mental Health* 1. DOI: 10.1016/j.ssmmh.2021.100008.

Brewster, M. 2021. "Pandemic early warning system has issued only a handful of alerts since start of 2020". Canadian Broadcasting Corporation, 7 October. www.cbc.ca/news/politics/pandemic-early-warning-covid-gphin-phac-1.6201910.

Brown, H. et al. 2012. "Our land, our language: connecting dispossession and health equity in an indigenous context". *Canadian Journal of Nursing Research* 44 (2): 44–63.

Brown, S. et al. 2019. "Health care experiences and birth outcomes: results of an Aboriginal birth cohort". *Women and Birth* 32 (5): 404–11.

Brown, T. 2011. "'Vulnerability is universal': considering the place of 'security' and 'vulnerability' within contemporary global health discourse". *Social Science & Medicine* 72 (3): 319–26.

Brown, T. 2018. "Global health geographies". In *Health Geographies: A Critical Introduction*, T. Brown et al. (eds), 251–70. Chichester: Wiley-Blackwell.

Brown, T. & G. Moon 2012. "Commentary: geography and global health". *Geographical Journal* 178 (1): 13–17.

Brown, T. & S. Taylor 2018. "Global health geographies". In *Routledge Handbook of Health Geography*, V. Crooks, G. Andrews & J. Pearce (eds), 14–19. Abingdon: Routledge.

Brown, T. et al. (eds) 2010. *A Companion to Health and Medical Geography*. Chichester: Wiley-Blackwell.

Brown, T. et al. (eds) 2018. *Health Geographies: A Critical Introduction*. Chichester: Wiley-Blackwell.

Brown, T., M. Cueto & E. Fee 2006. "The World Health Organization and the translation from 'international' to 'global' public health". In *Medicine at the Border: Disease, Globalization and Security, 1850 to the Present*, A. Bashford (ed.), 76–94. Basingstoke: Palgrave Macmillan.

Bruckner, B. et al. 2022. "Impacts of poverty alleviation on national and global carbon emissions". *Nature Sustainability* 5: 311–20.

Bryant, R. (ed.) 2015. *International Handbook of Political Ecology*. Cheltenham: Edward Elgar.

Bryson, J. et al. 2021. "Seasonality, climate change, and food security during pregnancy among indigenous and non-indigenous women in rural Uganda: implications for maternal-infant health". *PLoS ONE* 16 (3). DOI: 10.1371/journal.pone.0247198.

Burnett, K. et al. 2020. "Indigenous peoples, settler colonialism, and access to health care in rural and northern Ontario". *Health & Place* 66: DOI: 10.1016/j.healthplace.2020.102445.

Burns, J., A. Tomita & C. Lund 2017. "Income inequality widens the existing income-related disparity in depression risk in post-apartheid South Africa: evidence from a nationally representative panel study". *Health & Place* 45: 10–16.

Burstein, R. et al. 2019. "Mapping 123 million neonatal, infant and child deaths between 2000 and 2017". *Nature* 574: 353–8.

Busfield, J. 2015. "The globalization of the pharmaceutical industry". In *Globalization and Health*, J. Hanefeld (ed.), 127–42. Maidenhead: Open University Press.

Büyüm, A. et al. 2020. "Decolonising global health: if not now, when?". *BMJ Global Health* 5 (8). DOI: 10.1136/bmjgh-2020-003394.

Cai, H., Y. Chen & Q. Gong 2016. "Polluting thy neighbor: unintended consequences of China's pollution reduction mandates". *Journal of Environmental Economics and Management* 76: 86–104.

Calistri, P. et al. 2013. "The components of 'One World – One Health' approach". *Transboundary and Emerging Diseases* 60 (Supp. 2): 4–13.

Camfield, L., G. Crivello & M. Woodhead 2009. "Wellbeing research in developing countries: reviewing the role of qualitative methods". *Social Indicators Research* 90: 5–31.

Campbell, I. 2016. "Integrated management in the Mekong river basin". *Ecohydrology & Hydrobiology* 16 (4): 255–62.

REFERENCES

Carlson, C. et al. 2022. "Climate change increases cross-species viral transmission risk". *Nature* 607: 555–62.

Carlson, L. et al. 2015. "Petroleum pipeline explosions in sub-Saharan Africa: a comprehensive systematic review of the academic and lay literature". *Burns* 41 (3): 497–501.

Carlsten, C. et al. 2020. "Personal strategies to minimise effects of air pollution on respiratory health: advice for providers, patients and the public". *European Respiratory Journal* 55 (6). DOI: 10.1183/13993003.02056-2019.

Carrasco-Torrontegui, A. et al. 2020. "Climate change, food sovereignty, and ancestral farming technologies in the Andes". *Current Developments in Nutrition* 5 (Supp. 4): 54–60.

Carvalho, S., M. de Avalar Figueiredo Mafra Magalhães & R. de Andrade Medronho 2017. "Analysis of the spatial distribution of dengue cases in the city of Rio de Janeiro, 2011 and 2012". *Revista de Saúde Pública* 51. DOI: 10.11606/s1518-8787.2017051006239.

Castro, M. et al. 2019. "Brazil's unified health system: the first 30 years and prospects for the future". *The Lancet* 394: 345–56.

CDC 2014. "Update: Ebola virus disease outbreak – west Africa, October 2014". 31 October. www.cdc.gov/mmwr/preview/mmwrhtml/mm6343a3.htm?s_cid=mm6343a3_w.

Cernev, T. 2022. "Global catastrophic risk and planetary boundaries: the relationship to global targets and disaster risk reduction", contributing paper. Geneva: United Nations Office for Disaster Risk Reduction.

CIHR 2021. "Global health 3.0: CIHR's framework for action on global health research, 2021–2026". Ottawa: CIHR.

Chakraborty, J. & P. Basu 2021. "Air quality and environmental injustice in India: connecting particulate pollution to social disadvantages". *International Journal of Environmental Research and Public Health* 18 (1). DOI: 10.3390/ijerph18010304.

Chakravorty Spivak, G. 1988. "Can the subaltern speak?". In *Marxism and the Interpretation of Culture*, C. Nelson & L. Grossberg (eds), 271–313. Basingstoke: Macmillan.

Chatterjee, P. 2017. "The health system in India: the underserved majority". *The Lancet* 390: 2426–7.

Chaudhuri, M. et al. 2021. "Decolonising global health: beyond 'reformative' roadmaps and towards decolonial thought". *BMJ Global Health* 6 (7). DOI: 10.1136/bmjgh-2021-006371.

Chen, S. et al. 2019. "Current situation and progress toward the 2030 health-related Sustainable Development Goals in China: a systematic analysis". *PLoS Medicine* 16 (11). DOI: 10.1371/journal.pmed.1002975.

Choudhury, M. et al. 2018. "Cutaneous malignancy due to arsenicosis in Bangladesh: 12-year study in tertiary level hospital". *BioMed Research International*. DOI: 10.1155/2018/4678362.

Christakis, N. 2020. *Apollo's Arrow: The Profound and Enduring Impact of Coronavirus on the Way We Live*. New York: Little, Brown.

Clapp, J. 2002. "What the pollution havens debate overlooks". *Global Environmental Politics* 2 (2): 11–19.

Cliff, A. et al. 2009. *Infectious Diseases: A Geographical Analysis: Emergence and Re-Emergence*. Oxford: Oxford University Press.

Connell, J. & M. Walton-Roberts 2016. "What about the workers? The missing geographies of healthcare". *Progress in Human Geography* 40 (2): 158–76.

Coughlin, S. & A. Szema 2019. "Burn pits exposure and chronic respiratory illnesses among Iraq and Afghanistan veterans". *Journal of Environment and Health Sciences* 5 (1): 13–14.

Corburn, J. & A. Sverdlik 2017. "Slum upgrading and health equity". *International Journal of Environmental Research and Public Health* 14 (4). DOI: 10.3390/ijerph14040342.

Correa Massa, K. & A. Chiavegatto Filho 2021. "Income inequality and self-reported health among older adults in Brazil". *Journal of Applied Gerontology* 40 (2): 152–61.
Corsi, D. *et al.* 2009. "Gender inequity and age-appropriate immunization coverage in India from 1992 to 2006". *BMC International Health Human Rights* 9 (Supp. 1). DOI: 10.1186/1472-698X-9-S1-S3.
Cousins, T. *et al.* 2021. "The changing climates of global health". *BMJ Global Health* 6 (3). DOI: 10.1136/bmjgh-2021-005442.
COVID-19 Excess Mortality Collaborators 2022. "Estimating excess mortality due to the COVID-19 pandemic: a systematic analysis of COVID-19-related mortality, 2020–21". *The Lancet* 399: 1513–36.
CPHA 2015. "Global change and public health: addressing the ecological determinants of health", discussion paper. Ottawa: CPHA.
Crabtree, J. 2018. "Can India break out of the inequality trap?". Milken Institute, 31 July.
Craddock, S. 2017. "Making ties through making drugs: partnerships for tuberculosis drug and vaccine development". In *Global Health and Geographical Imaginaries*, C. Herrick & D. Reubi (eds), 76–93. Abingdon: Routledge.
Craggs, R. 2019. "Decolonising *The Geographical Tradition*". *Transactions of the Institute of British Geographers* 44 (3): 444–6.
Crighton, E., H. Gordon & C. Barakat-Haddad 2018. "Environmental health inequities: from global to local contexts". In *Routledge Handbook of Health Geography*, V. Crooks, G. Andrews & J. Pearce (eds), 37–44. Abingdon: Routledge.
Crocetti, A. *et al.* 2022. "The commercial determinants of Indigenous health and well-being: a systematic scoping review". *BMJ Global Health* 7 (11). DOI: 10.1136/bmjgh-2022-010366.
Crooks, V., G. Andrews & J. Pearce (eds) 2018. *Routledge Handbook of Health Geography*. Abingdon: Routledge.
Crowter, A. 2022. "Refugees in Wales: 'Thank God I came here after our home was bombed'". BBC News, 24 March. www.bbc.co.uk/news/uk-wales-60820974.
Cunsolo, A. *et al.* 2020. "You can never replace the caribou: Inuit experiences of ecological grief from caribou declines". *American Imago* 77 (1): 31–59.
Curtis, S. & K. Oven 2012. "Geographies of health and climate change". *Progress in Human Geography* 36 (5): 654–66.
Cyranoski, D. 2017. "Bat cave solves mystery of deadly SARS virus – and suggests new outbreak could occur". *Nature* 552: 15–16.
Da Rocha, D. *et al.* 2018. "The map of conflicts related to environmental injustice and health in Brazil". *Sustainability Science* 13: 709–19.
Dadari, I. *et al.* 2021. "Pro-equity immunization and health systems strengthening strategies in select GAVI-supported countries". *Vaccine* 39 (17): 2434–44.
Dahlgren, G. & M. Whitehead 2021. "The Dahlgren–Whitehead model of health determinants: 30 years on and still chasing rainbows". *Public Health* 199: 20–4.
Dai, Q. *et al.* 2020. "Severe dioxin-like compound (DLC) contamination in e-waste recycling areas: an under-recognized threat to local health". *Environment International* 139: DOI: 10.1016/j.envint.2020.105731.
Daly, T. 2019. "Populism, public law, and democratic decay in Brazil: understanding the rise of Jair Bolsonaro". Ramat Gan, Israel: College of Law and Business. https://clb.ac.il/wp-content/uploads/2018/12/Daly_Populism-Public-Law-Dem-Dec-Brazil_LEHR.pdf.
Daoud, N. *et al.* 2012. "Internal displacement and health among the Palestinian minority in Israel". *Social Science & Medicine* 74 (8): 1163–71.

REFERENCES

Davies, S. 2008. "Securitizing infectious disease". *International Affairs* 84 (2): 295–313.

Davies, S. 2021. "Antimicrobial resistance is the silent pandemic growing in the shadows". iNews, 27 May. https://inews.co.uk/news/health/dame-sally-davies-antimicrobial-resistance-is-the-silent-pandemic-growing-in-the-shadows-1023878.

Davies, S. *et al*. 2019. "Why it must be a feminist global health agenda". *The Lancet* 393: 601–3.

Davies, T. 2022. "Slow violence and toxic geographies: 'out of sight' to whom?". *Environment and Planning C: Politics and Space* 40 (2): 409–27.

Davies, T., A. Isakjee & S. Dhesi 2017. "Violent inaction: the necropolitical experience of refugees in Europe". *Antipode* 49 (5): 1263–84.

Davies, T., A. Isakjee & J. Obradovic-Wochnik 2022. "Epistemic borderwork: violent pushbacks, refugees, and the politics of knowledge at the EU border". *Annals of the American Association of Geographers* 113 (1): 169–88.

Davis, A. & J. Sharp 2020. "Rethinking One Health: emergent human, animal and environmental assemblages". *Social Science & Medicine* 258. DOI: 10.1016/j.socscimed.2020.113093.

Davis, M. 2006. *The Monster at Our Door: The Global Threat of Avian Influenza*. London: Macmillan.

De, S. *et al*. 2020. "Chronic respiratory morbidity in the Bhopal gas disaster cohorts: a time-trend analysis of cross-sectional data (1986–2016)". *Public Health* 186: 20–7.

De Lacy-Vawdon, C. & C. Livingstone 2020. "Defining the commercial determinants of health: a systematic review". *BMC Public Health* 20. DOI: 10.1186/s12889-020-09126-1.

De Leeuw, S. *et al*. 2012. "With reserves: colonial geographies and First Nations health". *Annals of the American Association of Geographers* 102 (5): 904–11.

De Oliveira Andrade, R. 2020. "Covid-19 is causing the collapse of Brazil's national health service". *BMJ* 370. DOI: 10.1136/bmj.m3032.

Dean, L. *et al*. 2019. "Neglected tropical disease as a 'biographical disruption': listening to the narratives of affected persons to develop integrated people centred care in Liberia". *PLoS Neglected Tropical Diseases* 13: DOI: 10.1371/journal.pntd.0007710.

Dion, M., P. AbdelMalik & A. Mawudeku 2015 "Big Data and the Global Public Health Intelligence Network (GPHIN)". *Canada Communicable Disease Report* 41 (9): 209–14.

Doctors Without Borders 2015. "Fighting Ebola across borders in Guinea and Sierra Leone". 16 April. www.doctorswithoutborders.org/latest/fighting-ebola-across-borders-guinea-and-sierra-leone.

Dorling, D. 2007. "Anamorphosis: the geography of physicians, and mortality". *International Journal of Epidemiology* 36 (4): 745–50.

Dorling, D. 2013. *Unequal Health: The Scandal of Our Times*. Bristol: Policy Press.

Dorling, D., R. Mitchell & J. Pearce 2007. "The global impact of income inequality on health by age: an observational study". *British Medical Journal* 335: 873–5.

Dos Anjos Luis, A. & P. Cabral 2016. "Geographic accessibility to primary healthcare centers in Mozambique". *International Journal of Equity in Health* 15. DOI: 10.1186/s12939-016-0455-0.

Doyle, T. & M. Risely (eds) 2008. *Crucible for Survival: Environmental Security and Justice in the Indian Ocean Region*. New Brunswick, NJ: Rutgers University Press.

Duff, C. 2018. "After posthumanism: health geographies of networks and assemblages". In *Routledge Handbook of Health Geography*, V. Crooks, G. Andrews & J. Pearce (eds), 137–43. Abingdon: Routledge.

Easterly, W. 2015. "The trouble with the Sustainable Development Goals". *Current History* 114: 322–4.

Ecks, S. & S. Basu 2019. "The unlicensed lives of antidepressants in India: generic drugs, unqualified practitioners, and floating prescriptions". *Transcultural Psychiatry* 46 (1): 86–106.

REFERENCES

Ecks, S. & I. Harper 2013. "Public–private mixes: the market for anti-tuberculosis drugs in India". In *When People Come First: Critical Studies in Global Health*, J. Biehl & A. Petryna (eds), 252–75. Princeton, NJ: Princeton University Press.

ECPAT 2022. "New data obtained from the Home Office shows only 2% of child victims of trafficking are given Discretionary Leave to Remain in the UK", press release. 17 February. www.ecpat.org.uk/news/new-data-obtained-from-the-home-office-shows-only-2-of-child-victims-of-trafficking-are-given-discretionary-leave-to-remain-in-the-uk.

Elbe, S. 2010. *Security and Global Health*. Cambridge: Polity Press.

Ellis-Petersen, H. 2022. "Covid-19: India accused of trying to delay WHO revision of death toll". *The Guardian*, 18 April. www.theguardian.com/world/2022/apr/18/covid-19-india-accused-of-attempting-to-delay-who-revision-of-death-toll.

Eni, R. *et al.* 2021. "Decolonizing health in Canada: a Manitoba first nation perspective". *International Journal of Equity and Health* 20 (1). DOI: 10.1186/s12939-021-01539-7.

Erdenee, O. *et al.* 2020. "Distribution of midwives in Mongolia: a secondary data analysis". *Midwifery* 86. DOI: 10.1016/j.midw.2020.102704.

ESA 2021. "Air pollution returning to pre-COVID levels". 15 March. www.esa.int/Applications/Observing_the_Earth/Copernicus/Sentinel-5P/Air_pollution_returning_to_pre-COVID_levels.

Escobar, A. 2020. *Pluriversal Politics: The Real and the Possible*. Durham, NC: Duke University Press.

Esquivel, V. 2016. "Power and the Sustainable Development Goals: a feminist analysis". *Gender & Development* 24 (1): 9–23.

Esri 2020. "Mapping coronavirus, responsibly". 25 February. www.esri.com/arcgis-blog/products/product/mapping/mapping-coronavirus-responsibly.

Euromedmonitor 2021. "New report: 91% of Gaza children suffer from PTSD after the Israeli attack". 2 July. https://euromedmonitor.org/en/article/4497.

Evans, J. 2018. "Decentering geographies of health: the challenge of post-structuralism". In *Routledge Handbook of Health Geography*, V. Crooks, G. Andrews & J. Pearce (eds), 131–6. Abingdon: Routledge.

Faizan, M. & R. Thakur 2019. "Association between solid cooking fuels and respiratory disease across socio-demographic groups in India". *Journal of Health and Pollution* 9 (23). DOI: 10.5696/2156-9614-9.23.190911.

Farbotko, C. 2019. "Climate change displacement: towards ontological security". In *Dealing with Climate Change on Small Islands: Towards Effective and Sustainable Adaptation?*, C. Klöck & M. Fink (eds), 251–66. Göttingen: Göttingen University Press.

Farmer, P. 2005. *Pathologies of Power: Health, Human Rights, and the New War on the Poor*. Berkeley, CA: University of California Press.

Farmer, P. 2020. *Fevers, Feuds, and Diamonds: Ebola and the Ravages of History*. New York: Farrar, Strauss & Giroux.

Farmer, P. *et al.* 2013. "Global health priorities for the early twenty-first century". In *Reimagining Global Health: An Introduction*, P. Farmer *et al.* (eds), 302–39. Berkeley, CA: University of California Press.

Fattah, K. & P. Walters 2020. "'A good place for the poor!' Counternarratives to territorial stigmatisation from two informal settlements in Dhaka". *Social Inclusion* 8 (1): 55–75.

Faye, C. *et al.* 2020. "Large and persistent subnational inequalities in reproductive, maternal, newborn and child health intervention coverage in sub-Saharan Africa". *BMJ Global Health* 5 (1). DOI: 10.1136/bmjgh-2019-002232.

Faye, J. & Y. Braun 2022. "Soil and human health: understanding agricultural and socio-environmental risk and resilience in the age of climate change". *Health & Place* 77. DOI: 10.1016/j.healthplace.2022.102799.

REFERENCES

Feinstein, A. & I. Choonara 2020. "Arms sales and child health". *BMJ Paediatrics Open* 4 (1). DOI: 10.1136/bmjpo-2020-000809.

Feldbaum, H. *et al.* 2006. "Global health and national security: the need for critical engagement". *Medicine, Conflict and Survival* 22 (3): 192–8.

Ferdowsian, H. 2021. "Ecological justice and the right to health: an introduction". *Health and Human Rights Journal* 23 (2): 1–5.

Fernandez, B. 2018. "Health inequities faced by Ethiopian migrant domestic workers in Lebanon". *Health & Place* 50: 154–61.

Ferguson, R. & Z. Jamal 2002. "A health-based case against Canadian arms transfers to Saudi Arabia". *Health and Human Rights* 22 (2): 243–55.

Ferretti, F. 2020. "History and philosophy of geography I: decolonising the discipline, diversifying archives and historicising radicalism". *Progress in Human Geography* 44 (6): 1161–71.

Ferring, D. & H. Hausermann 2019. "The political ecology of landscape change, malaria, and cumulative vulnerability in central Ghana's gold mining country". *Annals of the American Association of Geographers* 109 (4): 1074–91.

Ferris, E. & K. Donato 2020. *Refugees, Migration and Global Governance: Negotiating the Global Compacts*. Abingdon: Routledge.

Fidler, D. 2004. *SARS, Governance and the Globalization of Disease*. London: Palgrave Macmillan.

Findling, M. *et al.* 2019. "Discrimination in the United States: experiences of Native Americans". *Health Services Research* 54 (Supp. 2): 1431–41.

Fisher, J. *et al.* 2021. "Perceived biodiversity, sound, naturalness and safety enhance the restorative quality and wellbeing benefits of green and blue space in a neotropical city". *Science of the Total Environment* 755. DOI: 10.1016/j.scitotenv.2020.143095.

Fornace, K. *et al.* 2019. "Local human movement patterns and land use impact exposure to zoonotic malaria in Malaysian Borneo". *eLife* 8. DOI: 10.7554/eLife.47602.

Forster, T. *et al.* 2020. "Globalization and health equity: the impact of structural adjustment programs on developing countries". *Social Science & Medicine* 267. DOI: 10.1016/j.socscimed.2019.112496.

Frankenberg, E. *et al.* 2008. "Mental health in Sumatra after the tsunami". *American Journal of Public Health* 98 (9): 1671–7.

Frenkel, S. & J. Western 1988. "Pretext or prophylaxis? Racial segregation and malarial mosquitos in a British tropical colony: Sierra Leone". *Annals of the American Association of Geographers* 78 (2): 211–28.

Frost, I. *et al.* 2019. "Global geographic trends in antimicrobial resistance: the role of international travel". *Journal of Travel Medicine* 26 (8). DOI: 10.1093/jtm/taz036.

Fuller, R. *et al.* 2022. "Pollution and health: a progress update". *Lancet Planetary Health* 6 (6): e535–e547.

Galisson, M. 2020. "Deadly crossings and the militarisation of Britain's borders". In *Deadly Crossings and the Militarisation of Britain's Borders*, IRR, 6–12. London: IRR.

Galvão de Araújo, J. *et al.* 2012. "Origin and evolution of dengue virus type 3 in Brazil". *PLoS Neglected Tropical Diseases* 6 (9). DOI: 10.1371/journal.pntd.0001784.

Gandy, M. 2008. "Deadly alliances: death, disease, and the global politics of public health". In *Networked Disease: Emerging Infections in the Global City*, H. Ali & R. Keil (eds), 172–85. Chichester: Wiley-Blackwell.

Gaspar, R. *et al.* 2021. "Income inequality and non-communicable disease mortality and morbidity in Brazil states: a longitudinal analysis 2002–2017". *Lancet Regional Health – Americas* 2. DOI: 10.1016/j.lana.2021.100042.

Gates Foundation 2022. "The future of progress: the 2022 goalkeepers report". Seattle: Gates Foundation.
Gates Foundation n.d.a. "Foundation fact sheet". www.gatesfoundation.org/about/foundation-fact-sheet.
Gates Foundation n.d.b. "Water, sanitation and hygiene". www.gatesfoundation.org/our-work/programs/global-growth-and-opportunity/water-sanitation-and-hygiene.
Gatrell, A. 2011. *Mobilities and Health*. Farnham: Ashgate.
Gatrell, A. & S. Elliott 2015. *Geographies of Health: An Introduction*, 3rd edn. Chichester: Wiley-Blackwell.
Gatrell, P. 2019. *The Unsettling of Europe: How Migration Reshaped a Continent*. New York: Basic Books.
Gautier, L. *et al.* 2020. "What links can be made from narratives of migration and self-perceived health? A qualitative study with Haitian migrants settling in Quebec after the 2010 Haiti earthquake". *Journal of Migration and Health* 1/2. DOI: 10.1016/j.jmh.2020.100017.
Gautier, L. *et al.* 2022. "Rethinking development interventions through the lens of decoloniality in sub-Saharan Africa: the case of global health". *Global Public Health* 17 (2): 180–93.
GBD 2019 Demographics Collaborators 2020. "Global age-sex-specific fertility, mortality, healthy life expectancy (HALE), and population estimates in 204 countries and territories, 1950–2019: a comprehensive demographic analysis for the Global Burden of Disease study 2019". *The Lancet* 396: 1160–203.
GBD 2019 Adolescent Mortality Collaborators 2021. "Global, regional, and national mortality among young people aged 10–24 years, 1950–2019: a systematic analysis for the Global Burden of Disease study 2019". *The Lancet* 398: 1593–618.
Gee, S. & M. Skovdal 2017. "Navigating 'riskscapes': the experiences of international health care workers responding to the Ebola outbreak in west Africa". *Health & Place* 45: 173–80.
Gezie, L. *et al.* 2021. "Exploring factors that contribute to human trafficking in Ethiopia: a socio-ecological perspective". *Globalization and Health* 17 (1). DOI: 10.1186/s12992-021-00725-0.
Gibb, R. *et al.* 2020a. "Zoonotic host diversity increases in human-dominated ecosystems". *Nature* 584: 398–402.
Gibb, R. *et al.* 2020b. "Ecosystem perspectives are needed to manage zoonotic risks in a changing climate". *British Medical Journal* 371. DOI: 10.1136/bmj.m3389.
Gill, N., J. Caletrío & V. Mason 2011. "Introduction: mobilities and forced migration". *Mobilities* 6 (3): 301–16.
Gill, S. & S. Benatar 2017. "History, structure and agency in global health governance". *International Journal of Health Policy and Management* 6 (4): 237–41.
Gillam, C. & A. Charles 2019. "Community wellbeing: the impacts of inequality, racism and environment on a Brazilian coastal slum". *World Development Perspectives* 13: 18–24.
Gillespie, S. *et al.* 2020. "Residential mobility, mental health, and community violence exposure among Somali refugees and immigrants in North America". *Health & Place* 65. DOI: 10.1016/j.healthplace.2020.102419.
Giordano, C. *et al.* 2009. Building partnerships: conversations with Native Americans about mental health needs and community strengths. Sacramento, CA: University of California – Davis Center for Reducing Health Disparities.
Glasgow, S. & T. Schrecker 2016. "The double burden of neoliberalism? Noncommunicable disease policies and the global political economy of risk". *Health & Place* 39: 204–11.
Glick Schiller, N. & N. Salazar 2013. "Regimes of mobility across the world". *Journal of Ethnic and Migration Studies* 39 (2): 183–200.
Gordon, S. *et al.* 2014. "Respiratory risks from household air pollution in low and middle income countries". *Lancet Respiratory Medicine* 2 (10): 823–60.

REFERENCES

Gostin, L., S. Moon & B. Meier 2020. "Reimagining global health governance in the age of COVID-19". *American Journal of Public Health* 110 (11): 1615–19.

Gracey, M. & M. King 2009. "Indigenous health part 1: determinants and disease patterns". *The Lancet* 374: 65–75.

Graham, S. 2015. "Life support: the political ecology of urban air". *City* 19 (2/3): 192–215.

Greenhough, B. 2018. "Health and medical tourism". In *Health Geographies: A Critical Introduction*, T. Brown *et al.* (eds), 234–50. Chichester: Wiley-Blackwell.

Grifferty, G. *et al.* 2021. "Vulnerabilities to and the socioeconomic and psychosocial impacts of the leishmaniases: a review". *Research and Reports in Tropical Medicine* 12: 135–51.

Griffiths, K. *et al.* 2021. "Delineating and analyzing locality-level determinants of cholera, Haiti". *Emerging Infectious Diseases* 27 (1): 170–81.

Griffiths, M. 2022. "The geontological time-spaces of late modern war". *Progress in Human Geography* 46 (2): 282–98.

Grove, N. & A. Zwi 2006. "Our health and theirs: forced migration, othering, and public health". *Social Science & Medicine* 62 (8): 1931–42.

Guardian, The 2019. "Oil tanker explosion kills Nigerians collecting leaking fuel". 12 January. www.theguardian.com/world/2019/jan/12/oil-tanker-explosion-kills-nigerians-collecting-leaking-fuel.

Gupta, P. & E. Rodary 2017. "Opening-up Mozambique: histories of the present". *African Studies* 76 (2): 179–87.

Hagopian, A. *et al.* 2010. "Trends in childhood leukemia in Basrah, Iraq, 1993–2007". *American Journal of Public Health* 100 (6): 1081–7.

HAI (Health Alliance International) n.d. "Our model". www.healthallianceinternational.org/our-model.

Halaimzai, Z. 2021. "'We tried to be joyful enough to deserve our new lives': what it's really like to be a refugee in Britain". *The Guardian*, 20 July. www.theguardian.com/world/2021/jul/20/new-lives-refugee-britain-afghanistan-asylum-uk-taliban.

Hall, B. *et al.* 2021. "Fast food restaurant density and weight status: a spatial analysis among Filipina migrant workers in Macao (SAR), People's Republic of China". *Social Science & Medicine* 269. DOI: 10.1016/j.socscimed.2020.113192.

Hanna-Attisha. M. *et al.* 2016. "Elevated blood lead levels in children associated with the Flint drinking water crisis: a spatial analysis of risk and public health response". *American Journal of Public Health* 106 (2): 283–90.

Harpham, T. 2009. "Urban health in developing countries: what do we know and where do we go?". *Health & Place* 15 (1): 107–16.

Harris, M. & E. Carter 2019. "Muddying the waters: a political ecology of mosquito-borne disease in coastal Ecuador". *Health & Place* 57: 330–8.

Harvey, D. 2006. "The geographies of critical geography". *Transactions of the Institute of British Geographers* 31 (4): 409–12.

Hassan, M., P. Atkins & C. Dunn 2005. "Social implications of arsenic poisoning in Bangladesh". *Social Science & Medicine* 61 (10): 2201–11.

Hay, M., J. Skinner & A. Norton 2019. "Dam-induced displacement and resettlement: a literature review", FutureDAMS Working Paper 004. Manchester: University of Manchester.

Hayden, S. 2022. "Droughts in Somalia are partly our fault. We could at least let more migrants in". *The Guardian*, 27 April. www.theguardian.com/world/somalia/2022/apr/27/all.

Heimer, R. *et al.* 2022. "Body territory: mapping women's resistance to violence in the favelas of Mare, Rio de Janeiro", research report. London: King's College London.

Helldén, D. et al. 2021. "Climate change and child health: a scoping review and an expanded conceptual framework". *Lancet Planetary Health* 5 (3): e164–e175.

Helliwell, J. et al. (eds) 2020. *World Happiness Report 2020*. New York: Sustainable Development Solutions Network.

Hern, A. 2022. "Rohingya refugees welcome US decision to call Myanmar atrocities a genocide". *The Guardian*, 22 March. www.theguardian.com/world/2022/mar/22/rohingya-refugees-welcome-us-decision-to-call-myanmar-atrocities-a-genocide.

Herrick, C. 2016. "Global health, geographical contingency, and contingent geographies". *Annals of the American Association of Geographers* 106 (3): 672–87.

Herrick, C. 2017a. "The strategic geographies of global health partnerships". *Health & Place* 45: 152–9.

Herrick, C. 2017b. "When places come first: suffering, archetypal space and the problematic production of global health". *Transactions of the Institute of British Geographers* 42 (4): 530–43.

Herrick, C. & K. Bell 2022. "Concepts, disciplines and politics: on 'structural violence' and the 'social determinants of health'". *Critical Public Health* 32 (3): 295–308.

Herrick, C. & A. Brooks 2020. "Global health volunteering, the Ebola outbreak, and instrumental humanitarianisms in Sierra Leone". *Transactions of the Institute of British Geographers* 45 (2): 362–76.

Herrick, C. & D. Reubi (eds) 2017a. *Global Health and Geographical Imaginaries*. Abingdon: Routledge.

Herrick, C. & D. Reubi 2017b. "Introduction". In *Global Health and Geographical Imaginaries*, C. Herrick & D. Reubi (eds), ix–xxviii. Abingdon: Routledge.

Herrick, C. & D. Reubi 2021. "The future of the global noncommunicable disease agenda after Covid-19". *Health & Place* 71. DOI: 10.1016/j.healthplace.2021.102672.

Herrick, C., A. Kelly & J. Soulard 2022. "Humanitarian inversions: COVID-19 as crisis". *Transactions of the Institute of British Geographers* 47 (4): 850–65.

Herrick, C., O. Okpako & J. Millington 2021. "Unequal ecosystems of global health authorial expertise: decolonising noncommunicable disease". *Health & Place* 71. DOI: 10.1016/j.healthplace.2021.102670.

Herring, D. & S. Lockerbie 2010. "The coming plague of avian influenza". In *Plagues and Epidemics: Infected Spaces Past and Present*, D. Herring & A. Swedlund (eds), 179–91. Oxford: Berg.

Herring, D. & L. Sattenspiel 2007. "Social contexts, syndemics, and infectious disease in northern Aboriginal populations". *American Journal of Human Biology* 19 (2): 190–202.

Herring, D. & A. Swedlund (eds) 2010. *Plagues and Epidemics: Infected Spaces Past and Present*. Oxford: Berg.

Hinchliffe, S. 2017. "More than one world, more than one health: re-configuring inter-species health". In *Global Health and Geographical Imaginaries*, C. Herrick & D. Reubi (eds), 159–75. Abingdon: Routledge.

Hinchliffe, S. 2022. "Postcolonial global health, post-colony microbes and antimicrobial resistance". *Theory, Culture & Society* 39 (3): 145–68.

Hirsch, L. 2021. "Race and the spatialisation of risk during the 2013–2016 west African Ebola epidemic". *Health & Place* 67. DOI: 10.1016/j.healthplace.2020.102499.

Hochschild, A. 2019. *King Leopold's Ghost: A Story of Greed, Terror and Heroism in Colonial Africa*. London: Picador.

Holloway, L. 2004. "Donna Haraway". In *Key Thinkers on Space and Place*, P. Hubbard, R. Kitchin & G. Valentine (eds), 167–73. London: Sage.

Home Office 2022. "World first partnership to tackle global migration crisis". 14 April. www.gov.uk/government/news/world-first-partnership-to-tackle-global-migration-crisis.

REFERENCES

Hong, A. *et al.* 2021. "Neighbourhood green space and health disparities in the global South: evidence from Cali, Colombia". *Health & Place* 72. DOI: 10.1016/j.healthplace.2021.102690.

Hossain, A. *et al.* 2019. "Health risks of Rohingya children in Bangladesh: 2 years on". *The Lancet* 394: 1413–14.

Hotez, P. 2008. *Forgotten People, Forgotten Diseases: The Neglected Tropical Diseases and Their Impact on Global Health and Development*. Washington, DC: ASM Press.

Houweling, T. *et al.* 2016. "Socioeconomic inequalities in neglected tropical diseases: a systematic review". *PLoS Neglected Tropical Diseases* 10 (5): DOI: 10.1371/journal.pntd.0004546.

HRW 2016. "Bangladesh: 20 million drink arsenic-laced water". 6 April. www.hrw.org/news/2016/04/06/bangladesh-20-million-drink-arsenic-laced-water.

HRW 2021. "Brazil". In *World Report 2021: Events of 2020*, 105–16. New York: HRW.

Hunter, M. & L. Lawson 2020. *A Rough Cut Trade: Africa's Coloured-Gemstone Flows to Asia*. Geneva: Global Initiative against Transnational Organized Crime.

Huo, X. *et al.* 2007. "Elevated blood lead levels of children in Guiyu, an electronic waste recycling town in China". *Environmental Health Perspectives* 115 (7): 1113–17.

Hurtig, A.-K. & M. San Sebastián 2002. "Geographical differences in cancer incidence in the Amazon basin of Ecuador in relation to residence near oil fields". *International Journal of Epidemiology* 31 (5): 1021–37.

Hutchinson, H. 2010. "My heart is in two places: ontological security, emotions and the health of African refugee women in Tasmania". Unpublished PhD thesis, University of Tasmania. https://eprints.utas.edu.au/17681/2/hutchison_thesis.pdf.

Hyatt, R. 2006. "Military spending: global health threat or global public good?". In *Globalization and Health*, I. Kawachi & S. Wamala (eds), 311–29. Oxford: Oxford University Press.

ICC Research Foundation 2021. "The economic case for global vaccinations". Paris: ICC.

IDMC 2021a. *2021 Internal Displacement Index Report*. Geneva: IDMC.

IDMC 2021b. *GRID 2021: Internal Displacement in a Changing Climate*. Geneva: IDMC.

IHME 2020. "Latest GBD results: 2019". 15 October. www.healthdata.org/gbd/gbd-2019-resources.

IHME n.d. "Democratic Republic of the Congo". www.healthdata.org/democratic-republic-congo.

IMPEL 2020. "Surge in global e-waste, up 21% in 5 years, according to Global E-Waste Monitor 2020". 2 July. www.impel.eu/surge-in-global-e-waste-up-21-in-5-years-according-to-global-e-waste-monitor-2020.

India State-Level Disease Burden Initiative Air Pollution Collaborators 2021. "Health and economic impact of air pollution in the states of India: the Global Burden of Disease study 2019". *Lancet Planetary Health* 5 (1): e25–e38.

IPCC 2022. *Climate Change 2022: Impacts, Adaptation and Vulnerability: Contribution of Working Group II to the Sixth Assessment Report of the Intergovernmental Panel on Climate Change*. Cambridge: Cambridge University Press.

Iqbal, N. *et al.* 2018. "Girls' hidden penalty: analysis of gender inequality in child mortality with data from 195 countries". *BMJ Global Health* 3 (5). DOI: 10.1136/bmjgh-2018-001028.

Iranzo, A. 2021. "Sub-Saharan migrants 'in transit': intersections between mobility and immobility and the production of (in)securities". *Mobilities* 16 (5): 739–57.

IRR 2020. *Deadly Crossings and the Militarisation of Britain's Borders*. London: IRR.

Jabbour, S. *et al.* (eds) 2012. *Public Health in the Arab World*. Cambridge: Cambridge University Press.

Jackson, P. & A. Neely 2015. "Triangulating health: toward a practice of a political ecology of health". *Progress in Human Geography* 39 (1): 47–64.

Jang, S., Y. Ekyalongo & H. Kim 2021. "Systematic review of displacement and health impact from natural disasters in southeast Asia". *Disaster Medicine and Public Health Preparedness* 15 (1): 105–14.

Jepson, W. *et al.* 2017. "Advancing human capabilities for water security: a relational approach". *Water Security* 1: 46–52.

Jia, J. *et al.* 2020. "Population flow drives spatio-temporal distribution of COVID-19 in China". *Nature* 582: 389–94.

Jiang, B. *et al.* 2020. "Health impacts of environmental contamination of micro- and nanoplastics: a review". *Environmental Health and Preventive Medicine* 25 (1). DOI: 10.1186/s12199-020-00870-9.

Johnson, K. *et al.* 2010. "Association of sexual violence and human rights violations with physical and mental health in territories of the eastern Democratic Republic of the Congo". *Journal of the American Medical Association* 304 (5): 553–62.

Jones N. *et al.* 2020. "Water, sanitation and hygiene risk factors for the transmission of cholera in a changing climate: using a systematic review to develop a causal process diagram". *Journal of Water and Health* 18 (2): 145–58.

Jose, A. 2019. "India's regional disparity and its policy responses". *Journal of Public Affairs* 19 (4). DOI: 10.1002/pa.1933.

Jungari, S. *et al.* 2022. "Violence against women in urban slums of India: a review of two decades of research". *Global Public Health* 17 (1): 115–33.

Kalipeni, E., J. Ghosh & J. Oppong 2017. "Disease and illness, political ecology of". In *The International Encyclopaedia of Geography: People, the Earth, Environment and Technology*, D. Richardson *et al.* (eds), 15 vols. Chichester: Wiley-Blackwell. DOI: 10.1002/9781118786352.wbieg0540.

Kamradt-Scott, A. 2016. "WHO's to blame? The World Health Organization and the 2014 Ebola outbreak in west Africa". *Third World Quarterly* 37 (3): 401–18.

Kangmennaang, J. & S. Elliott 2019. "'Wellbeing is shown in our appearance, the food we eat, what we wear, and what we buy': embodying wellbeing in Ghana". *Health & Place* 55: 177–87.

Kakarla, S. *et al.* 2020. "Dengue situation in India: suitability and transmission potential model for present and projected climate change scenarios". *Science of the Total Environment* 739. DOI: 10.1016/j.scitotenv.2020.140336.

Kashwan, P. *et al.* 2020. "Planetary justice: prioritizing the poor in Earth system governance". *Earth System Governance* 6. DOI: 10.1016/j.esg.2020.100075.

Kawachi, I. & S. Wamala (eds) 2006a. *Globalization and Health*. Oxford: Oxford University Press.

Kawachi, I. & S. Wamala 2006b. "Poverty and inequality in a globalizing world". In *Globalization and Health*, I. Kawachi & S. Wamala (eds), 122–37. Oxford: Oxford University Press.

Kearns, R. 1993. "Place and health: towards a reformed medical geography". *The Professional Geographer* 45 (2): 139–47.

Kelman, I. *et al.* 2019. "Does climate change influence people's migration decisions in Maldives?". *Climatic Change* 153 (1/2): 285–99.

Kemp, C. & L. Rasbridge 2004. *Refugee and Immigrant Health: A Handbook for Health Professionals*. Cambridge: Cambridge University Press.

Kennes, M. 2021. "'In Israel, you're 60 times more likely to have a COVID vaccine than in Palestine'". MSF, 22 February. www.msf.org/stark-inequality-covid-19-vaccination-between-israel-and-palestine.

Kephart, J. *et al.* 2022. "City-level impact of extreme temperatures and mortality in Latin America". *Nature Medicine* 28 (8): 1700–5.

Kessel, A. 2011. *Air, the Environment, and Public Health*. Cambridge: Cambridge University Press.

REFERENCES

Khan, K. *et al.* 2009. "Spread of a novel influenza A (H1N1) virus via global airline transportation". *New England Journal of Medicine* 361 (2): 212–14.

Khan, T. *et al.* 2022. "How we classify countries and people – and why it matters". *BMJ Global Health* 7 (6). DOI: 10.1136/bmjgh-2022-009704.

Khosla, R., P. Allotey & S. Gruskin 2020. "Global health and human rights for a postpandemic world". *BMJ Global Health* 5 (8). DOI: 10.1136/bmjgh-2020-003548.

Kickbusch, I. 2005. "Tackling the political determinants of global health". *British Medical Journal* 331: 246–7.

Kickbusch, I. 2006. "The need for a European strategy on global health". *Scandinavian Journal of Public Health* 34 (6): 561–5.

Kickbusch, I. 2015. "The political determinants of health – 10 years on". *British Medical Journal* 350. DOI: 10.1136/bmj.h81.

Kickbusch, I., L. Allen & C. Franz 2016. "The commercial determinants of health". *Lancet Global Health* 4 (12): e895–e896.

Kim, M. 2019. "The effects of transboundary air pollution from China on ambient air quality in South Korea". *Heliyon* 5 (12). DOI: 10.1016/j.heliyon.2019.e02953.

Kim, R. *et al.* 2019. "Micro-geographic targeting for precision public policy: analysis of child sex ratio across 587,043 census villages in India, 2011". *Health & Place* 57: 92–100.

King, B. 2017. *States of Disease: Political Environments and Human Health*. Oakland, CA: University of California Press.

King, M., A. Smith & M. Gracey 2009. "Indigenous health part 2: the underlying causes of the health gap". *The Lancet* 374: 76–85.

King, N. 2002. "Security, disease, commerce: ideologies of postcolonial global health". *Social Studies of Science* 32 (5/6): 763–89.

Kirby, P. & A. Lora-Wainwright 2015. "Exporting harm, scavenging value: transnational circuits of e-waste between Japan, China and beyond". *Area* 47 (1): 40–7.

Klinenberg, E. 2002. *Heat Wave: Social Autopsy of a Disaster in Chicago*. Chicago: University of Chicago Press.

Koplan, J. *et al.* 2009. "Towards a common definition of global health". *The Lancet* 373: 1993–5.

Kounnavong, S. *et al.* 2017. "Malaria elimination in Lao PDR: the challenges associated with population mobility". *Infectious Diseases of Poverty* 6 (1). DOI: 10.1186/s40249-017-0283-5.

Krause-Jackson, F. 2011. "Clinton chastises China on internet, African 'new colonialism'". Bloomberg UK, 11 June. www.bloomberg.com/news/articles/2011-06-11/clinton-chastises-china-on-internet-african-new-colonialism-?leadSource=uverify%20wall.

Krishnamoorthy, Y. *et al.* 2018. "Emerging public health threat of e-waste management: global and Indian perspective". *Review of Environmental Health* 33 (4): 321–9.

Kumar, A. *et al.* 2020. "Success story of Dharavi against COVID-19". *British Medical Journal* 370. DOI: 10.1136/bmj.m2817.

Kumar, P., C. Paton & D. Korigia 2016. "I've got 99 problems but a phone ain't one: electronic and mobile health in low and middle income countries". *Archives of Disease in Childhood* 101 (10): 974–9.

Kushner, J. 2020. "'It became part of life': how Haiti curbed cholera". *The Guardian*, 16 March. www.theguardian.com/global-development/2020/mar/16/it-became-part-of-life-how-haiti-curbed-cholera.

Labonté, R. & T. Schrecker 2007. "Globalization and social determinants of health: introduction and methodological background". *Global Health* 3 (5). DOI: 10.1186/1744-8603-3-5.

REFERENCES

Laffert, B. & D. Sala 2021. "Conflict and climate change collide: why northeast Syria is running dry". The New Humanitarian, 20 December. www.thenewhumanitarian.org/news-feature/2021/12/20/conflict-climate-change-why-northeast-Syria-is-running-dry.

Lai, S. et al. 2018. "Seasonal and interannual risks of dengue introduction from south-east Asia into China, 2005–2015". PLoS Neglected Tropical Diseases 12 (11). DOI: 10.1371/journal.pntd.0006743.

Lakoff, A. 2017. Unprepared: Global Health in a Time of Emergency. Oakland, CA: University of California Press.

Lana, R. et al. 2017. "The introduction of dengue follows transportation infrastructure changes in the state of Acre, Brazil: a network-based analysis". PLoS Neglected Tropical Diseases 11 (11). DOI: 10.1371/journal.pntd.0006070.

Lana, R. et al. 2021. "The top 1%: quantifying the unequal distribution of malaria in Brazil". Malaria Journal 20. DOI: 10.1186/s12936-021-03614-4.

Lea, J. 2018. "Non-representational theory and health geographies". In Routledge Handbook of Health Geography, V. Crooks, G. Andrews & J. Pearce (eds), 144–52. Abingdon: Routledge.

Learmonth, A. 1988. Disease Ecology: An Introduction. Oxford: Blackwell.

Lee, K. 2015. "Introduction to the global economy". In Globalization and Health, J. Hanefeld (ed.), 26–48. Maidenhead: Open University Press.

Lencucha, R. & S. Neupane 2022. "The use, misuse and overuse of the 'low-income and middle-income countries' category". BMJ Global Health 7 (6). DOI: 10.1136/bmjgh-2022-009067.

Lerer, L. & T. Scudder 1999. "Health impacts of large dams". Environmental Impact Assessment Review 19 (2): 113–23.

Leuenberger, A. et al. 2021. "Health impacts of industrial mining on surrounding communities: local perspectives from three sub-Saharan African countries". PLoS ONE 16 (6). DOI: 10.1371/journal.pone.0252433.

Lewis, J., J. Hoover & D. MacKenzie 2017. "Mining and environmental health disparities in Native American communities". Current Environmental Health Reports 4 (2): 130–41.

Liao, Y. et al. 2010. "Spatial analysis of neural tube defects in a rural coal mining area". International Journal of Environmental Health Research 20 (6): 439–50.

Liberty 2022. "How does the new Policing Act affect my protest rights?". 29 June. www.libertyhumanrights.org.uk/advice_information/pcsc-policing-act-protest-rights.

Lilford, R. et al. 2019. "Because space matters: conceptual framework to help distinguish slum from non-slum urban areas". BMJ Global Health 4 (2). DOI: 10.1136/bmjgh-2018-001267.

Liu, L. et al. 2019. "National, regional, and state-level all-cause and cause-specific under-5 mortality in India in 2000–15: a systematic analysis with implications for the Sustainable Development Goals". Lancet Global Health 7 (6): e721–e734.

Liu, S. et al. 2020. "Revealing the impacts of transboundary pollution on PM2.5-related deaths in China". Environment International 134. DOI: 10.1016/j.envint.2019.105323.

Liverani, M. 2015. "Globalization and infectious diseases". In Globalization and Health, J. Hanefeld (ed.), 155–64. Maidenhead: Open University Press.

Local Burden of Disease Diarrhoea Collaborators 2020. "Mapping geographical inequalities in oral rehydration therapy coverage in low-income and middle-income countries, 2000–17". Lancet Global Health 8 (8): e1038–e1060.

Local Burden of Disease Vaccine Coverage Collaborators 2021. "Mapping routine measles vaccination in low- and middle-income countries". Nature 589: 415–19.

REFERENCES

Local Burden of Disease WaSH Collaborators 2020. "Mapping geographical inequalities in access to drinking water and sanitation facilities in low-income and middle-income countries, 2000–17". *Lancet Global Health* 8 (9): e1162–e1185.

Lovell, A., U. Read & C. Lang 2019. "Genealogies and anthropologies of global mental health". *Culture, Medicine and Psychiatry* 43 (4): 519–47.

Lowe, R. *et al.* 2020. "Emerging arboviruses in the urbanized Amazon rainforest". *British Medical Journal* 371. DOI: 10.1136/bmj.m4385.

Luginaah, I. & R. Bezner-Kerr (eds) 2015. *Geographies of Health and Development*. Farnham: Ashgate.

Luo, W. & Y. Xie 2020. "Economic growth, income inequality and life expectancy in China". *Social Science & Medicine* 256. DOI: 10.1016/j.socscimed.2020.113046.

Macassa, G. *et al.* 2012. "Geographic differentials in mortality of children in Mozambique: their implications for achievement of Millennium Development Goal 4". *Journal of Health, Population and Nutrition* 30 (3): 331–45.

McCoy, D. & K. Kinyua 2012. "Allocating scarce resources strategically – an evaluation and discussion of the Global Fund's pattern of disbursements". *PLoS ONE* 7 (5). DOI: 10.1371/journal.pone.0034749.

McCreesh, N., G. Nikulin & M. Booth 2015. "Predicting the effects of climate change on *Schistosoma mansoni* transmission in eastern Africa". *Parasites & Vectors* 8 (4). DOI: 10.1186/s13071-014-0617-0.

McDonald, A. *et al.* 2017. "Invisible wounds: the impact of six years of war on the mental health of Syria's children". London: Save the Children.

McGeachan, C. & C. Philo 2017. "Occupying space: mental health geography and global directions". In *The Palgrave Handbook of Sociocultural Perspectives on Global Mental Health*, R. White *et al.* (eds), 31–50. Basingstoke: Palgrave Macmillan.

McGoey, L. 2016. *No Such Thing as a Free Gift: The Gates Foundation and the Price of Philanthropy*. London: Verso.

Machado, C. & G. Silva 2019. "Political struggles for a universal health system in Brazil: successes and limits in the reduction of inequalities". *Globalization and Health* 15 (Supp. 1). DOI: 10.1186/s12992-019-0523-5.

Machado, D. *et al.* 2020. "Monitoring the progress of health-related sustainable development goals (SDGs) in Brazilian states using the Global Burden of Disease indicators". *Population Health Metrics* 18 (Supp. 1). DOI: 10.1186/s12963-020-00207-2.

McInnes, C., K. Lee & J. Youde (eds) 2020. *The Oxford Handbook of Global Health Politics*. Oxford: Oxford University Press.

MacKay, D. *et al.* 2022. "Exceeding 1.5°C global warming could trigger multiple climate tipping points". *Science* 377. DOI: 10.1126/science.abn7950.

McLafferty, S., F. Wang & J. Butler 2011. "Rural–urban inequalities in late-stage breast cancer: spatial and social dimensions of risk and access". *Environment and Planning B: Planning and Design* 38 (4): 726–40.

Madge, C. 1998. "Therapeutic landscapes of the Jola, The Gambia, west Africa". *Health & Place* 4: 293–311.

Mahajan, M. 2018. "Philanthropy and the nation-state in global health: the Gates Foundation in India". *Global Public Health* 13 (10): 1357–68.

Mahmudi-Azer, S. 2006. "Arms trade and its impact on global health". *Theoretical Medicine and Bioethics* 27 (1): 81–93.

Maju, M. *et al.* 2019. "A colonial legacy of HIV/AIDS, NTD, and STI super-syndemics: eugenicist foreign aid and intertwined health burdens in Nigeria". *Global Public Health* 14 (9): 1221–40.

Mandavilli, A. 2018. "The world's worst industrial disaster is still unfolding". *The Atlantic*, 10 July. www.theatlantic.com/science/archive/2018/07/the-worlds-worst-industrial-disaster-is-still-unfolding/560726.

Mappr n.d. "Provinces of the Democratic Republic of the Congo". www.mappr.co/counties/congo-provinces.

Markham, L. 2022. "'A disaster waiting to happen': who was really responsible for the fire at Moria refugee camp?". *The Guardian*, 21 April. www.theguardian.com/world/2022/apr/21/disaster-waiting-to-happen-moria-refugee-camp-fire-greece-lesbos.

Marshall, W. 2021. "The political economy of Mozambique's 'faceless insurgency'". Global Risk Insights, 19 April. https://globalriskinsights.com/2021/04/the-political-economy-of-mozambiques-faceless-insurgency.

Martens, P. & L. Hall 2000. "Malaria on the move: human population movement and malaria transmission". *Emerging Infectious Disease* 6 (2): 103–9.

Martinez, R. *et al.* 2020. "Trends in premature avertable mortality from non-communicable diseases for 195 countries and territories, 1990–2017: a population-based study". *Lancet Global Health* 8 (4): e511–e523.

Massad, S. *et al.* 2011. "Health-related quality of life of Palestinian preschoolers in the Gaza Strip: a cross-sectional study". *BMC Public Health* 11. DOI: 10.1186/1471-2458-11-253.

Mayer, J. 1996. "The political ecology of disease as one new focus for medical geography". *Progress in Human Geography* 20 (4): 441–56.

Mayer, J. 2010. "Medical geography". In *A Companion to Health and Medical Geography*, T. Brown, S. McLafferty & G. Moon (eds), 33–54. Oxford: Blackwell.

Mayer, J. & M. Meade 1994. "A reformed medical geography reconsidered". *The Professional Geographer* 46 (1): 103–6.

Meehan, K. *et al.* 2020. "Geographies of insecure water access and the housing-water nexus in US cities". *Proceedings, National Academy of Sciences USA* 117 (46): 28700–7.

Mekong River Commission 2019. *State of the Basin Report 2018*. Vientiane, Laos: Mekong River Commission.

Meng, J. *et al.* 2018. "Origin and radiative forcing of black carbon aerosol: production and consumption perspectives". *Environmental Science & Technology* 52 (11): 6380–9.

Menon, J. *et al.* 2020. "What was right about Kerala's response to the COVID-19 pandemic?". *BMJ Global Health* 5 (7). DOI: 10.1136/bmjgh-2020-003212.

Merdian, H. *et al.* 2019. "Transnational child sexual abuse: outcomes from a roundtable discussion". *International Journal of Environmental Research and Public Health* 16 (2). DOI: 10.3390/ijerph16020243.

Messina, J. *et al.* 2016. "Mapping global environmental suitability for Zika virus". *eLife* 5. DOI: 10.7554/eLife.15272.

Met Office 2022. "Climate change making heatwaves more intense", press release. 18 May. www.metoffice.gov.uk/about-us/press-office/news/weather-and-climate/2022/southern-asian-heatwave-attribution-study-2022.

Meyers, T. & N. Hunt 2014. "The other global South". *The Lancet* 384: 1921–2.

Middleton, J. *et al.* 2020. "'We're people of the snow': weather, climate change, and Inuit mental wellness". *Social Science & Medicine* 262. DOI: 10.1016/j.socscimed.2020.113137.

Middleton, J. *et al.* 2021. "Temperature and place associations with Inuit mental health in the context of climate change". *Environmental Research* 198. DOI: 10.1016/j.envres.2021.111166.

REFERENCES

Milligan, C. 2018. "Home truths? A critical reflection on aging, care and the home". In *Routledge Handbook of Health Geography*, V. Crooks, G. Andrews & J. Pearce (eds), 230–6. Abingdon: Routledge.

Mills, C. & S. Fernando 2014. "Globalising mental health or pathologising the Global South? Mapping the ethics, theory and practice of Global Mental Health". *Disability and the Global South* 1 (2): 188–202.

Minca, C. et al. 2022. "Rethinking the biopolitical: borders, refugees, mobilities … ". *Environment and Planning C: Politics and Space* 40 (1): 3–30.

Mir, D. et al. 2021. "Recurrent dissemination of SARS-CoV-2 through the Uruguayan–Brazilian border". *Frontiers in Microbiology* 12. DOI: 10.3389/fmicb.2021.653986.

Mishra, S. 2015. "Putting value to human health in coal mining region of India". *Journal of Health Management* 17 (3): 339–55.

Mitchell, F. 2019. "Water (in)security and American Indian health: social and environmental justice implications for policy, practice, and research". *Public Health* 176: 98–105.

Mitlin, D. & D. Satterthwaite (eds) 2012. *Urban Poverty in the Global South: Scale and Nature*. Abingdon: Routledge.

Moolgavkar, S. et al. 2014. "Cancer mortality and quantitative oil production in the Amazon region of Ecuador, 1990–2010". *Cancer Causes and Control* 25: 59–72.

Moon, S. 2019. "Power in global governance: an expanded typology from global health". *Globalization and Health* 15 (Supp. 1). DOI: 10.1186/s12992-019-0515-5.

Moosa, S. et al. 2014. "The inverse primary care law in sub-Saharan Africa: a qualitative study of the views of migrant health workers". *British Journal of General Practice* 64: e321–e328.

Mora, C. 2022. "Over half of known human pathogenic diseases can be aggravated by climate change". *Nature Climate Change* 12 (9): 869–75.

Mordecai, E. et al. 2020. "Climate change could shift disease burden from malaria to arboviruses in Africa". *Lancet Planetary Health* 4 (9): e416–e423.

Morrice, E. & R. Colagiuri 2013. "Coal mining, social injustice and health: a universal conflict of power and priorities". *Health & Place* 19: 74–9.

Mostafanezhad, M. & O. Evrard 2021. "Chronopolitics of crisis: a historical political ecology of seasonal air pollution in northern Thailand". *Geoforum* 124: 400–8.

Mugabe, V. et al. 2021. "Natural disasters, population displacement and health emergencies: multiple public health threats in Mozambique". *BMJ Global Health* 6 (9). DOI: 10.1136/bmjgh-2021-006778.

Mukherjee, D. et al. 2017. "Park availability and major depression in individuals with chronic conditions: is there an association in urban India?". *Health & Place* 47: 54–62.

Mulligan, K., S. Elliott & C. Schuster-Wallace 2012. "The place of health and the health of place: Dengue fever and urban governance in Putrajaya, Malaysia". *Health & Place* 18 (3): 613–20.

Murray, C. et al. 2006. "Eight Americas: investigating mortality disparities across races, counties, and race-counties in the United States". *PLoS Medicine* 3 (9). DOI: 10.1371/journal.pmed.0030260.

Musolino, C. et al. 2020. "Global health activists' lessons on building social movements for Health for All". *International Journal of Equity in Health* 19. DOI: 10.1186/s12939-020-01232-1.

Nansai, K. et al. 2021. "Consumption in the G20 nations causes particulate air pollution resulting in two million premature deaths annually". *Nature Communications* 12. DOI: 10.1038/s41467-021-26348-y.

Narayan, R. 2006. "The role of the People's Health Movement in putting the social determinants of health on the global agenda". *Health Promotion Journal of Australia* 17 (3): 186–8.

Nature 2020. "Time to revise the Sustainable Development Goals". 583: 331–2.

Navarro, V. 1999. "Health and equity in the world in the era of 'globalization'". *International Journal of Health Services Research* 29 (2): 215–26.

Neely, A. & A. Nading 2017. "Global health from the outside: the promise of place-based research". *Health & Place* 45: 55–63.

Newbold, K. 2018. "Immigrant health: insights and implications". In *Routledge Handbook of Health Geography*, V. Crooks, G. Andrews & J. Pearce (eds), 202–8. Abingdon: Routledge.

New York Times 2021. "Covid vaccinations tracker". www.nytimes.com/interactive/2021/world/covid-vaccinations-tracker.html.

Nichols, C. & V. Del Casino 2021. "Towards an integrated political ecology of health and bodies". *Progress in Human Geography* 45 (4): 776–95.

Nicoson, C. 2021. "Towards climate resilient peace: an intersectional and degrowth approach". *Sustainability Science* 16: 1147–58.

Nixon, R. 2011. *Slow Violence and the Environmentalism of the Poor*. Cambridge, MA: Harvard University Press.

Nji, T. *et al.* 2021. "Eliminating onchocerciasis within the Meme river basin of Cameroon: a social-ecological approach to understanding everyday realities and health systems". *PLoS Neglected Tropical Diseases* 15. DOI: 10.1371/journal.pntd.0009433.

Noxolo, P. 2017. "Introduction: decolonising geographical knowledge in a colonised and re-colonising postcolonial world". *Area* 49 (3): 317–19.

Nti, A. *et al.* 2020. "Effect of particulate matter exposure on respiratory health of e-waste workers at Agbogbloshie, Accra, Ghana". *International Journal of Environmental Research and Public Health* 17 (9). DOI: 10.3390/ijerph17093042.

Nunbogu, A. & S. Elliott 2021. "Towards an integrated theoretical framework for understanding water insecurity and gender-based violence in low- and middle-income countries (LMICs)". *Health & Place* 71. DOI: 10.1016/j.healthplace.2021.102651.

Nunes, J. 2014. *Security, Emancipation and the Politics of Health: A New Theoretical Perspective*. Abingdon: Routledge.

Nyantakyi-Frimpong, H. 2021. "Climate change, women's workload in smallholder agriculture, and embodied political ecologies of undernutrition in northern Ghana". *Health & Place* 68. DOI: 10.1016/j.healthplace.2021.102536.

O'Callaghan-Gordo, C. *et al.* 2021. "Blood lead levels in indigenous peoples living close to oil extraction areas in the Peruvian Amazon". *Environment International* 154. DOI: 10.1016/j.envint.2021.106639.

O'Neill, K. 2019. *Waste*. Cambridge: Polity Press.

Okwaraji, Y. *et al.* 2012. "Effect of geographical access to health facilities on child mortality in rural Ethiopia: a community based cross sectional study". *PLoS ONE* 7 (3). DOI: 10.1371/journal.pone.0033564.

OPHI 2017. "Country briefing 2017: Mozambique". Oxford: OPHI. https://ophi.org.uk/wp-content/uploads/MOZ_2017.pdf.

Ormond, M. & M. Toyota 2018. "Rethinking care through transnational health and long-term care practices". In *Routledge Handbook of Health Geography*, V. Crooks, G. Andrews & J. Pearce (eds), 237–43. Abingdon: Routledge.

Ortner, S. 2016. "Dark anthropology and its others: theory since the eighties". *Journal of Ethnographic Theory* 6 (1): 47–73.

Oti, S. & J. Ncayiyana 2021. "Decolonising global health: where are the Southern voices?". *BMJ Global Health* 6 (7). DOI: 10.1136/bmjgh-2021-006576.

Ottawa CityNews 2021 "Brazil enjoys fun in the sun as COVID-19 deaths top 200,000". 8 January. https://ottawa.citynews.ca/world-news/brazil-enjoys-fun-in-the-sun-as-covid-19-deaths-top-200000-3243635.

REFERENCES

PAHO 1978. "Declaration of Alma-Ata". September. www3.paho.org/hq/index.php?option=com_content&view=article&id=13774:declaration-of-alma-ata&Itemid=0&lang=en#gsc.tab=0.

Pallister-Wilkins, P. 2020. "Hotspots and the geographies of humanitarianism". *Environment and Planning D: Society and Space* 38 (6): 991–1008.

Pallister-Wilkins, P. 2022. *Humanitarian Borders: Unequal Mobility and Saving Lives*. London: Verso.

Pandolfelli, L., J. Shandra & J. Tyagi 2014. "The International Monetary Fund, structural adjustment, and women's health: a cross-national analysis of maternal mortality in sub-Saharan Africa". *Sociological Quarterly* 55 (1): 119–42.

Papworth, A., M. Maslin & S. Randalls 2015. "Is climate change the greatest threat to global health?". *Geographical Journal* 181 (4): 413–22.

Partlow, J. & J. Warrick 2016. "A dangerous export: America's car-battery waste is making Mexican communities sick". *Washington Post*, 26 February. www.washingtonpost.com/sf/national/2016/02/26/a-dangerous-export-americas-car-battery-waste-is-making-mexican-communities-sick.

Patel, P., L. Kiapi & E. Gómez 2022. "Launching a new series on non-communicable prevention in humanitarian settings". *BMJ Global Health* 7. DOI: 10.1136/bmjgh-2022-009710.

Patel, V. *et al.* 2018a. "The *Lancet* Commission on global mental health and sustainable development". *The Lancet* 392: 1553–98.

Patel, V. *et al.* 2018b. "Income inequality and depression: a systematic review and meta-analysis of the association and a scoping review of mechanisms". *World Psychiatry* 17 (2): 76–89.

Patterson, D. 2022. "The right to health and the climate crisis: the vital role of civic space". *Health and Human Rights Journal* 23 (2): 109–20.

Pattisson, P. 2021. "'Like slave and master': DRC miners toil for 30p an hour to fuel electric cars". *The Guardian*, 8 November. www.theguardian.com/global-development/2021/nov/08/cobalt-drc-miners-toil-for-30p-an-hour-to-fuel-electric-cars.

Patton, G. *et al.* 2016. "Our future: a *Lancet* Commission on adolescent health and wellbeing". *The Lancet* 387: 2423–78.

Pearce, J., C. Tisch & R. Barnett 2008. "Have geographical inequalities in cause-specific mortality in New Zealand increased during the period 1980–2001?". *New Zealand Medical Journal* 121: 15–27.

Pécoud, A. 2021. "Narrating an ideal migration world? An analysis of the Global Compact for Safe, Orderly and Regular Migration". *Third World Quarterly* 42 (1): 16–33.

PEPFAR n.d. "Partnerships". www.state.gov/pepfar-partnerships/#:~:text=PEPFAR%20works%20closely%20with%20partner%20countries%20toward%20achieving,treatment%20work%20addressing%20key%20gaps%20in%20innovative%20ways.

Permanyer, I. 2013. "A critical assessment of the UNDP's Gender Inequality Index". *Feminist Economics* 19 (2): 1–32.

Pfeiffer, J. 2013. "The struggle for a public sector: PEPFAR in Mozambique". In *When People Come First: Critical Studies in Global Health*, J. Biehl & A. Petryna (eds), 166–81. Princeton, NJ: Princeton University Press.

Pfeiffer, J. *et al.* 2008. "Strengthening health systems in poor countries: a code of conduct for nongovernmental organizations". *American Journal of Public Health* 98 (12): 2134–40.

Physicians for Human Rights 2002. "Maternal mortality in Herat province, Afghanistan: the need to protect women's rights". Boston: Physicians for Human Rights.

PIH n.d.a. "Mission". www.pih.org/our-mission.

PIH n.d.b. "Building strong health systems". www.pih.org/our-approach.

Plamondon, K. & E. Bisung 2019. "The CCGHR Principles for Global Health Research: centering equity in research, knowledge translation, and practice". *Social Science & Medicine* 239. DOI: 10.1016/j.socscimed.2019.112530.

Pongou, R., J. Salomon & M. Ezzati 2006. "Health impacts of macroeconomic crises and policies: determinants of variation in childhood malnutrition trends in Cameroon". *International Journal of Epidemiology* 35 (3): 648–56.

Popa, E. 2020. "Mental health, normativity, and local knowledge in global perspective". *Studies in History and Philosophy of Biological and Biomedical Sciences* 84. DOI: 10.1016/j.shpsc.2020.101334.

Price-Smith, A. 2009. *Contagion and Chaos: Disease, Ecology and National Security in the Era of Globalization*. Cambridge, MA: MIT Press.

Puri, P. *et al*. 2020. "A cross-sectional study on selected child health outcomes in India: quantifying the spatial variations and identification of the parental risk factors". *Scientific Reports* 10. DOI: 10.1038/s41598-020-63210-5.

Puyvallée, A. & K. Storeng 2022. "COVAX, vaccine donations and the politics of global vaccine inequity". *Globalization and Health* 18 (1). DOI: 10.1186/s12992-022-00801-z.

Querejazu, A. 2016. "Encountering the pluriverse: looking for alternatives in other worlds". *Revista Brasileira de Política Internacional* 59 (2). DOI: 10.1590/0034-7329201600207.

Radcliffe, S. 2017. "Decolonising geographical knowledges". *Transactions of the Institute of British Geographers* 42 (3): 329–33.

Ramirez, M. *et al*. 2017. "Contamination by oil crude extraction: refinement and their effects on human health". *Environmental Pollution* 231: 415–25.

Rao, C. *et al*. 2020. "Premature adult mortality in India: what is the size of the matter?". *BMJ Global Health* 6 (6). DOI: 10.1136/bmjgh-2020-004451.

Rasella, D. *et al*. 2016. "Assessing the relevance of indicators in tracking social determinants and progress toward equitable population health in Brazil". *Global Health Action* 9 (Spec. 3). DOI: 10.3402/gha.v9.29042.

Reichenheim, M. *et al*. 2011. "Violence and injuries in Brazil: the effect, progress made, and challenges ahead". *The Lancet* 377: 1962–75.

Reidpath, D. & P. Allotey 2019. "The problem of 'trickle-down science' from the Global North to the Global South". *BMJ Global Health* 4 (4). DOI: 10.1136/bmjgh-2019-001719.

Reubi, D., C. Herrick & T. Brown 2016. "The politics of non-communicable diseases in the global South". *Health & Place* 39: 179–87.

Ricci, J. 2009. "Global health governance and the state: premature claims of a post-international framework". *Global Health Governance* 3 (1): 1–18.

Richardson, E., T. McGinnis & R. Frankfurter 2019. "Ebola and the narrative of mistrust". *BMJ Global Health* 4 (6): DOI: 10.1136/bmjgh-2019-001932.

Richmond, C. & K. Big-Canoe 2018. "The geographies of Indigenous health". In *Routledge Handbook of Health Geography*, V. Crooks, G. Andrews & J. Pearce (eds), 179–88. Abingdon: Routledge.

Richmond, C. *et al*. 2005. "The political ecology of health: perceptions of environment, economy, health and well-being among 'Namgis First Nation". *Health & Place* 11 (4): 349–65.

Ricketts, T. 2010. "Accessing health care". In *A Companion to Health and Medical Geography*, T. Brown, S. McLafferty & G. Moon (eds), 521–39. Oxford: Blackwell.

Rifkin, S. 2018. "Alma Ata after 40 years: primary health care and health for all – from consensus to complexity". *BMJ Global Health* 3 (Supp. 3). DOI: 10.1136/bmjgh-2018-001188.

REFERENCES

Riley, A. *et al.* 2020. "Systematic human rights violations, traumatic events, daily stressors and mental health of Rohingya refugees in Bangladesh". *Conflict and Health*. DOI: 10.1186/s13031-020-00306-9.

Ríos, V., E. Denova-Gutiérrez & S. Barquera 2022. "Association between living in municipalities with high crowding conditions and poverty and mortality from COVID-19 in Mexico". *PLoS ONE* 17 (2). DOI: 10.1371/journal.pone.0264137.

Ritchie, H. & M. Roser 2019. "Outdoor air pollution". Our World in Data, November. https://ourworldindata.org/outdoor-air-pollution.

Rocheleau, D., B. Thomas-Slayter & E. Wangari (eds) 1996. *Feminist Political Ecology: Global Issues and Local Experiences*. London: Routledge.

Rock, M., C. Degeling & G. Blue 2014. "Toward stronger theory in critical public health: insights from debates surrounding posthumanism". *Critical Public Health* 24 (3): 337–48.

Romanello, M. *et al.* 2021. "The 2021 report of the *Lancet Countdown* on health and climate change: code red for a healthy future". *The Lancet* 398: 1619–62.

Rosen, J. *et al.* 2021. "'Burnt by the scorching sun': climate-induced livelihood transformations, reproductive health, and fertility trajectories in drought-affected communities of Zambia". *BMC Public Health* 21 (1). DOI: 10.1186/s12889-021-11560-8.

Roy, A. 1999. *The Cost of Living*. London: Flamingo.

Ruiz-Cantero M. *et al.* 2019. "Governance commitment to reduce maternal mortality: a political determinant beyond the wealth of the countries". *Health & Place* 57: 313–20.

Rupasinghe, R., B. Chomel & B. Martínez-López 2022. "Climate change and zoonoses: a review of the current status, knowledge gaps, and future trends". *Acta Tropica* 226. DOI: 10.1016/j.actatropica.2021.106225.

Rushton, S. & J. Youde (eds) 2015. *Routledge Handbook of Global Health Security*. Abingdon: Routledge.

Ryan, H., L. Girion & S. Glover 2016. "OxyContin goes global – 'we're only just getting started'". *Los Angeles Times*, 18 December. www.latimes.com/projects/la-me-oxycontin-part3.

Ryan, S., C. Lippi & F. Zermoglio 2020. "Shifting transmission risk for malaria in Africa with climate change: a framework for planning and intervention". *Malaria Journal* 19. DOI: 10.1186/s12936-020-03224-6.

Sacks, J. *et al.* 2022. "The Lancet Commission on lessons for the future from the COVID-19 pandemic". *The Lancet* 400: 1224–80.

Sakue-Collins, Y. 2021. "(Un)doing development: a postcolonial enquiry of the agenda and agency of NGOs in Africa". *Third World Quarterly* 42 (5): 976–95.

Saleh, S. *et al.* 2021. "Exploring smoke: an ethnographic study of air pollution in rural Malawi". *BMJ Global Health* 6 (6). DOI: 10.1136/bmjgh-2021-004970.

Salih, A. *et al.* 2020. "Climate change and locust outbreak in East Africa". *Nature Climate Change* 10 (10): 584–5.

San Martín, W. & N. Wood 2022. "Pluralising planetary justice beyond the North–South divide: recentring procedural, epistemic, and recognition-based justice in earth-systems governance". *Environmental Science and Policy* 128: 256–63.

San Sebastián, M. *et al.* 2001. "Exposures and cancer incidence near oil fields in the Amazon basin of Ecuador". *Occupational and Environmental Medicine* 58 (8): 517–22.

Sassen, S. 1996. *Losing Control? Sovereignty in an Age of Globalization*. New York: Columbia University Press.

Savioli, L. *et al.* 2017. "Building a global schistosomiasis alliance: an opportunity to join forces to fight inequality and rural poverty". *Infectious Diseases of Poverty* 6 (1). DOI: 10.1186/s40249-017-0280-8.

Schoeps, A. *et al.* 2011. "The effect of distance to health-care facilities on childhood mortality in rural Burkina Faso". *American Journal of Epidemiology* 173 (5): 492–8.

Schrecker, T. 2020. "Towards a critical political economy of global health". In *The Oxford Handbook of Global Health Politics*, C. McInnes, K. Lee & J. Youde (eds), 469–90. Oxford: Oxford University Press.

Schrecker, T., A.-E. Birn & M. Aguilera 2018. "How extractive industries affect health: political economy underpinnings and pathways". *Health & Place* 52: 135–47.

Schur, N. *et al.* 2013. "Spatially explicit Schistosoma infection risk in eastern Africa using Bayesian geostatistical modelling". *Acta Tropica* 128 (2): 365–77.

Schuster, B. 2009. "Gaps in the Silk Road: an analysis of population health disparities in the Xinjiang Uyghur Autonomous Region of China". *China Quarterly* 198: 433–41.

Secretariat of the Basel Convention n.d. "The Convention". www.basel.int/TheConvention/Overview/tabid/1271/Default.aspx.

Selby, J. *et al.* 2017. "Climate change and the Syrian civil war revisited". *Political Geography* 60: 232–44.

Sen, A. 1990. "More than 100 million women are missing". *New York Review of Books* 37 (20): 61–66.

Severson, M. & D. Collins 2018. "Well-being in health geography: conceptualizations, contributions and questions". In *Routledge Handbook of Health Geography*, V. Crooks, G. Andrews & J. Pearce (eds), 124–30. Abingdon: Routledge.

Shah, R. 2020. "A Western delusion: narratives surrounding neocolonialism in Africa". Oxford Political Review, 23 April. http://oxfordpoliticalreview.com/amp/2020/04/23/a-western-delusion-narratives-surrounding-neocolonialism-in-africa.

Shantz, E. & S. Elliott 2021. "From social determinants to social epigenetics: health geographies of chronic disease". *Health & Place* 69. DOI: 10.1016/j.healthplace.2021.102561.

Sharara, S. & S. Kanj 2014. "War and infectious diseases: challenges of the Syrian civil war". *PLoS Pathogens* 10 (11). DOI: 10.1371/journal.ppat.1004438.

Sharma, R. 2021. "Migrant crisis: more than 150 people died crossing Channel in last 5 years and the total could be far higher". I News, 26 November. https://inews.co.uk/news/migrant-crisis-deaths-crossing-channel-figures-total-higher-warning-1318804.

Sheather, J. *et al.* 2016. "A Médecins Sans Frontières ethics framework for humanitarian innovation". *PLoS Medicine* 13 (9). DOI: 10.1371/journal.pmed.1002111.

Sheller, M. 2013. "The islanding effect: post-disaster mobility systems and humanitarian logistics in Haiti". *Cultural Geographies* 20 (2): 185–204.

Sheller, M. 2015. "Sociology after the mobilities turn". In *The Routledge Handbook of Mobilities*, P. Adey *et al.* (eds), 45–54. Abingdon: Routledge.

Shetty, S. 2014. "Thirty years on from Bhopal disaster: still fighting for justice". Amnesty International, 2 December. www.amnesty.org/en/latest/news/2014/12/thirty-years-bhopal-disaster-still-fighting-justice.

Shibata, T. *et al.* 2015. "Life in a landfill slum, children's health, and the Millennium Development Goals". *Science of the Total Environment* 536: 408–18.

Shiffman, J. & Y. Shawar 2020. "Strengthening accountability of the global health metrics enterprise". *The Lancet* 395: 1452–6.

Shrivastava, P. 1992. *Bhopal: Anatomy of a Crisis*, 2nd edn. London: Paul Chapman.

REFERENCES

Silva, J. 2008. "A multilevel analysis of agricultural trade and socioeconomic inequality in rural Mozambique". *The Professional Geographer* 60 (2): 174–89.

Singer, M. 2010. "Ecosyndemics: global warming and the coming plagues of the twenty-first century". In *Plagues and Epidemics: Infected Spaces Past and Present*, D. Herring & A. Swedlund (eds), 21–38. Oxford: Berg.

Smit, W. *et al*. 2016. "Making unhealthy places: the built environment and non-communicable diseases in Khayelitsha, Cape Town". *Health & Place* 39: 196–203.

Smith, J. 2015. *Epic Measures: One Doctor, Seven Billion Patients*. New York: Harper Wave.

Sneddon, C. & C. Fox 2016. "Rethinking transboundary waters: a critical hydropolitics of the Mekong basin". *Political Geography* 25 (2): 181–202.

Soares, G. *et al*. 2022. "Disparities in excess mortality between Indigenous and non-Indigenous Brazilians in 2020: measuring the effects of the COVID-19 pandemic". *Journal of Racial and Ethnic Health Disparities* 9 (6): 2227–36.

Sochas, L. 2020. "The predictive power of health system environments: a novel approach for explaining inequalities in access to maternal healthcare". *BMJ Global Health* 4 (Supp. 5). DOI: 10.1136/bmjgh-2019-002139.

Sohel, N. *et al*. 2010. "Spatial patterns of fetal loss and infant death in an arsenic-affected area in Bangladesh". *International Journal of Health Geographics* 9. DOI: 10.1186/1476-072X-9-53.

Solomonian, L. & E. Di Ruggiero 2021. "The critical intersection of environmental and social justice: a commentary". *Globalization and Health* 17 (1). DOI: 10.1186/s12992-021-00686-4.

Sorichetta, A. *et al*. 2016. "Mapping internal connectivity through human migration in malaria endemic countries". *Scientific Data* 3. DOI: 10.1038/sdata.2016.66.

Sotero, M. 2006. "A conceptual model of historical trauma: implications for public health practice and research". *Journal of Health Disparities Research and Practice* 1 (1): 93–108.

Souza, A. *et al*. 2018. "Geography of microcephaly in the Zika era: a study of newborn distribution and socio-environmental indicators in Recife, Brazil, 2015–2016". *Public Health Reports* 133 (4): 461–71.

Sovacool, B. 2019. "The precarious political economy of cobalt: balancing prosperity, poverty, and brutality in artisanal and industrial mining in the Democratic Republic of the Congo". *Extractive Industries and Society* 6 (3): 915–39.

Sparke, M. & D. Anguelov 2012. "H1N1, globalization and the epidemiology of inequality". *Health & Place* 18 (4): 726–36.

Sparke, M. & D. Anguelov 2020. "Contextualising coronavirus geographically". *Transactions of the Institute of British Geographers* 45 (3): 498–508.

Sparke, M. & O. Williams 2022. "Neoliberal disease: COVID-19, co-pathogenesis and global health insecurities". *Environment and Planning A: Economy and Space* 54 (1): 15–32.

Sperling, J. & J. Decker 2007. "The therapeutic landscapes of the Kaqchikel of San Lucas Tolimán, Guatemala". In *Therapeutic Landscapes*, A. Williams (ed.), 233–53. Farnham: Ashgate.

Spicer, N. 2015. "Non-government actors in global health". In *Globalization and Health*, J. Hanefeld (ed.), 89–99. Maidenhead: Open University Press.

Spiegelhalter, D. & A. Masters 2021. *Covid by Numbers: Making Sense of the Pandemic by Data*. London: Pelican Books.

Sridhar, D. 2022. *Preventable: How a Pandemic Changed the World and How to Stop the Next One*. London: Penguin Books.

Steffen, W. et al. 2015. "Sustainability: planetary boundaries: guiding human development on a changing planet". *Science* 347. DOI: 10.1126/science.1259855.

Stewart, J. 2021. "Long-term air pollution linked to greater risk of COVID-19 hospitalisation". Imperial College London, 6 September. www.imperial.ac.uk/news/229233/long-term-pollution-linked-greater-risk-covid-19.

Stjernborg, V., A. Wretstrand & M. Tesfahuney 2015. "Everyday life mobilities of older persons: a case study of ageing in a suburban landscape in Sweden". *Mobilities* 10 (3): 383–401.

Stöckl, H. et al. 2021. "Human trafficking and violence: findings from the largest global dataset of trafficking survivors". *Journal of Migration and Health* 4. DOI: 10.1016/j.jmh.2021.100073.

Stoeva, P. 2015. "Global health and security". In *Globalization and Health*, J. Hanefeld (ed.), 186–99. Maidenhead: Open University Press.

Strano E. et al. 2018. "Mapping road network communities for guiding disease surveillance and control strategies". *Scientific Reports* 8. DOI: 10.1038/s41598-018-22969-4.

Stuckler, D. et al. 2017. "Austerity and health: the impact in the UK and Europe". *European Journal of Public Health* 27 (Supp. 4): 18–21.

Sultana, F. 2006. "Gendered waters, poisoned wells: political ecology of the arsenic crisis in Bangladesh". In *Fluid Bonds: Views on Gender and Water*, K. Lahiri-Dutt (ed.), 362–86. Kolkata: Stree.

Sultana, F. 2011. "Suffering for water, suffering from water: emotional geographies of resource access, control and conflict". *Geoforum* 42 (2): 163–72.

Sultana, F. 2012. "Producing contaminated citizens: toward a nature–society geography of health and well-being". *Annals of the American Association of Geographers* 102 (5): 1165–72.

Sultana, F. 2022. "The unbearable heaviness of climate coloniality". *Political Geography* 99. DOI: 10.1016/j.polgeo.2022.102638.

Sultana, F. & A. Loftus (eds) 2019. *Water Politics: Governance, Justice, and the Right to Water*. Abingdon: Routledge.

Suri, A. et al. 2013. "Values and global health". In *Reimagining Global Health: An Introduction*, P. Farmer et al. (eds), 245–86. Berkeley, CA: University of California Press.

Swain, R. 2017. "A critical analysis of the Sustainable Development Goals". In *Handbook of Sustainability Science and Research*, W. Filho (ed.), 341–55. Cham, Switzerland: Springer Nature.

Swartz, A. et al. 2018. "Toxic layering through three disciplinary lenses: childhood poisoning and street pesticide use in Cape Town, South Africa". *Medical Humanities* 44 (4): 247–52.

Swinburn, B. et al. 2019. "The global syndemic of obesity, undernutrition, and climate change: the Lancet Commission report". *The Lancet* 393: 791–846.

Swyngedouw, E. 2009. "The political economy and political ecology of the hydro-social cycle". *Journal of Contemporary Water Research & Education* 142 (1): 56–60.

Szwarcwald, C. et al. 2011. "Health inequalities in Rio de Janeiro, Brazil: lower healthy life expectancy in socioeconomically disadvantaged areas". *American Journal of Public Health* 101 (3): 517–23.

Szwarcwald, C. et al. 2016. "Inequalities in healthy life expectancy by Brazilian geographic regions: findings from the National Health Survey, 2013". *International Journal for Equity in Health* 15. DOI: 10.1186/s12939-016-0432-7.

Tallman, P. et al. 2022. "Ecosyndemics: the potential synergistic health impacts of highways and dams in the Amazon". *Social Science & Medicine* 295. DOI: 10.1016/j.socscimed.2020.113037.

Tao, L. et al. 2021. "A geographically weighted regression model for health improvement: insights from the extension of life expectancy in China". *Applied Sciences* 11 (5). DOI: 10.3390/app11052022.

REFERENCES

Tatem, A., D. Rogers & S. Hay 2006. "Estimating the malaria risk of African mosquito movement by air travel". *Malaria Journal* 5. DOI: 10.1186/1475-2875-5-57.

Taylor, D. 2021. "UK waste company fined £1.5m for exporting household waste". *The Guardian*, 30 July. www.theguardian.com/environment/2021/jul/30/uk-waste-firm-fined-15m-for-exporting-household-waste.

Taylor, D. 2022. "'I thought the UK was a good country': Sudan massacre refugee faces removal to Rwanda". *The Guardian*, 18 May. www.theguardian.com/politics/2022/may/18/i-thought-the-uk-was-a-good-country-man-faces-removal-to-rwanda.

Taylor, M. 2014. *The Political Ecology of Climate Change Adaptation: Livelihoods, Agrarian Change and the Conflicts of Development*. London: Earthscan.

Tegegne, T. *et al*. 2019. "Antenatal care use in Ethiopia: a spatial and multilevel analysis". *BMC Pregnancy Childbirth* 19 (1). DOI: 10.1186/s12884-019-2550-x.

Terry, G. 2009. "No climate justice without gender justice: an overview of the issues". *Gender and Development* 17 (1): 5–18.

Theobald, S. *et al*. 2017. "20 years of gender mainstreaming in health: lessons and reflections for the neglected tropical diseases community". *BMJ Global Health* 2 (4). DOI: 10.1136/bmjgh-2017-000512.

Thomson, M., A. Kentikelenis & T. Stubbs 2017. "Structural adjustment programmes adversely affect vulnerable populations: a systematic-narrative review of their effect on child and maternal health". *Public Health Reviews* 38 (13). DOI: 10.1186/s40985-017-0059-2.

Thow, A. & C. Hawkes 2009. "The implications of trade liberalization for diet and health: a case study from Central America". *Globalization and Health* 5. DOI: 10.1186/1744-8603-5-5.

Tilley, L. *et al*. 2022. "Race and climate change: towards anti-racist ecologies". *Politics*. DOI: 10.1177/02633957221127166.

Tudor Hart, J. 1971. "The inverse care law". *The Lancet* 297: 407–12.

Tukuitonga, C. & P. Vivili 2021. "Climate effects on health in small islands developing states". *Lancet Planetary Health* 5 (2): e69–e70.

Uitermark, J. & J. Tieleman 2021. "From fragmentation to integration and back again: the politics of water infrastructure in Accra's peripheral neighbourhoods". *Transactions of the Institute of British Geographers* 46 (2): 347–62.

UN n.d.a. "The 17 goals". https://sdgs.un.org/goals.

UN n.d.b. "Genocide". www.un.org/en/genocideprevention/genocide.shtml.

UNDP n.d. "Gender Inequality Index (GII)". https://hdr.undp.org/data-center/thematic-composite-indices/gender-inequality-index#/indicies/GII.

UNEP 2021. *Adaptation Gap Report 2021: The Gathering Storm: Adapting to Climate Change in a Post-Pandemic World*. Nairobi: UNEP.

UNHCR 2022. *Global Trends: Forced Displacements in 2021*. Geneva: UNHCR.

UNICEF 2021. "Children bear brunt of violence in Gaza". 21 May. www.unicef.org/stories/children-bear-brunt-violence-gaza.

UNODC 2021a. "Awah Francisca Mbuli – Cameroon". www.unodc.org/unodc/en/endht/2021/survivor-stories---francisca-awah.html.

UNODC 2021b. "Marcela Loaiza – Colombia/United States". www.unodc.org/unodc/en/endht/2021/survivor-stories---marcela-loaiza.html.

UN Refugees & Migrants 2018. "Global Compact for Migration". https://refugeesmigrants.un.org/migration-compact.

UN Statistics Division 2017. "Goal 11: make cities and human settlements inclusive, safe, resilient and sustainable". https://unstats.un.org/sdgs/report/2017/goal-11.

UN Water 2017. "What is water security?". May. www.unwater.org/app/uploads/2017/05/unwater_poster_Oct2013.pdf.

UN Water 2021. *Summary Progress Update 2021: SDG 6 – Water and Sanitation for All*. Geneva: UN Water. www.unwater.org/app/uploads/2021/12/SDG-6-Summary-Progress-Update-2021_Version-July-2021a.pdf.

Urry, J. 2003. *Global Complexity*. Cambridge: Polity Press.

USA for UNHCR 2023. "Syria refugee crisis explained". 14 March. www.unrefugees.org/news/syria-refugee-crisis-explained.

Utazi, C. *et al.* 2022. "Assessing the characteristics of un- and under-vaccinated children in low- and middle-income countries: a multi-level cross-sectional study". *PLoS Global Public Health* 2 (4). DOI: 10.1371/journal.pgph.0000244.

Valoi, E. 2016. "The blood rubies of Montepuez". *Foreign Policy*, 3 May. https://foreignpolicy.com/2016/05/03/the-blood-rubies-of-montepuez-mozambique-gemfields-illegal-mining.

Van Bortel, T. *et al.* 2019. "Perceived stressors and coping mechanisms of female migrant domestic workers in Singapore". *PLoS ONE* 14 (3). DOI: 10.1371/journal.pone.0210717.

Van Brusselen, D. *et al.* 2020. "Metal mining and birth defects: a case-control study in Lubumbashi, Democratic Republic of the Congo". *Lancet Planetary Health* 4 (4): e158–e167.

Van Loon, J. 2002. *Risk and Technological Culture: Towards a Sociology of Virulence*. London: Routledge.

Van Wagner, E. 2008. "Toward a dialectical understanding of networked disease in the global city: vulnerability, connectivity, topologies". In *Networked Disease: Emerging Infections in the Global City*, H. Ali & R. Keil (eds), 13–26. Chichester: Wiley-Blackwell.

Varkey, S., E. Kandpal & S. Neelsen 2022. "Why addressing inequality must be central to pandemic preparedness". *BMJ Global Health* 7 (9). DOI: 10.1136/bmjgh-2022-010453.

Vearey, J. *et al.* 2010. "Urban health in Johannesburg: the importance of place in understanding intra-urban inequalities in a context of migration and HIV". *Health & Place* 16 (4): 694–702.

Véron, R. 2006. "Remaking urban environments: the political ecology of air pollution in Delhi". *Environment and Planning (A): Economy and Space* 38 (11): 2093–109.

Veronese, G. *et al.* 2020. "Spatial agency as a source of resistance and resilience among Palestinian children living in Dheisheh refugee camp, Palestine". *Health & Place* 62. DOI: 10.1016/j.healthplace.2020.102304.

Veseth, M. 2005. *Globaloney: Unraveling the Myths of Globalization*. Lanham, MD: Rowman & Littlefield.

Victora, C. *et al.* 2019. "Analyses of inequalities in RMNCH: rising to the challenge of the SDGs". *BMJ Global Health* 4 (Supp. 4). DOI: 10.1136/bmjgh-2018-001295.

Villarroel, M., T. Clarke & T. Norris 2020. "Health of American Indian and Alaska Native adults, by urbanization level: United States, 2014–2018", Data Brief 372. Hyattsville, MD: National Center for Health Statistics.

Vincens, N., M. Emmelin & M. Stafström 2018. "The interplay of contextual layers: a multilevel analysis of income distribution, neighborhood infrastructure, socioeconomic position and self-rated health in Brazil". *Health & Place* 55: 155–62.

Vincens, N. & M. Stafström 2015. "Income inequality, economic growth and stroke mortality in Brazil: longitudinal and regional analysis 2002–2009". *PLoS ONE* 10 (9). DOI: 10.1371/journal.pone.0137332.

REFERENCES

Voce, A., L. Cecco & C. Michael 2021. "'Cultural genocide': the shameful history of Canada's residential schools – mapped". *The Guardian*, 6 September. www.theguardian.com/world/ng-interactive/2021/sep/06/canada-residential-schools-indigenous-children-cultural-genocide-map.

Von Grebmer, K. et al. 2022. *Global Hunger Index 2022: Food Systems Transformation and Local Governance*. Bonn: Deutsche Welthungerhilfe and Concern Worldwide.

Wagner, Z. et al. 2018. "Armed conflict and child mortality in Africa". *The Lancet* 392: 857–65.

Walker, G. 2009. "Globalizing environmental justice: the geography and politics of frame contextualization and evolution". *Global Social Policy* 9 (3): 335–82.

Wallace, R. 2016. *Big Farms Make Big Flu: Dispatches on Infectious Disease, Agribusiness, and the Nature of Science*. New York: Monthly Review Press.

Walters, K. et al. 2011. "Bodies don't just tell stories, they tell histories: embodiment of historical trauma among American Indians and Native Alaskans". *Du Bois Review* 8 (1): 179–89.

Wang, H. 2021. "Why the Navajo Nation was hit so hard by coronavirus: understanding the disproportionate impact of the COVID-19 pandemic". *Applied Geography* 134. DOI: 10.1016/j.apgeog.2021.102526.

Wang, Y. et al. 2016. "Under-5 mortality in 2851 Chinese counties, 1996–2012: a subnational assessment of achieving MDG 4 goals in China". *The Lancet* 387: 273–83.

Ward, J. & R. Viner 2017. "The impact of income inequality and national wealth on child and adolescent mortality in low and middle-income countries". *BMC Public Health* 17. DOI: 10.1186/s12889-017-4310-z.

WaterAid 2020. "Cholera remains a deadly threat for the world's poor as progress towards elimination lags". 9 June. www.wateraid.org/uk/media/cholera-remains-threat.

Webber, F. 2020. "Introduction". In *Deadly Crossings and the Militarisation of Britain's Borders*, IRR, 3–5. London: IRR.

Weber, L. & S. Pickering 2011. *Globalization and Borders: Death at the Global Frontier*. London: Palgrave Macmillan.

WEF 2021. "This is how COVID-19 has affected world hunger". 13 July. www.weforum.org/agenda/2021/07/world-hunger-malnutrition-un-report-covid-coronavirus-pandemic-sdgs.

Weir, L. & E. Mykhalovsky 2006. "The geopolitics of global public health surveillance in the twenty-first century". In *Medicine at the Border: Disease, Globalization and Security, 1850 to the Present*, A. Bashford (ed.), 240–63. Basingstoke: Palgrave Macmillan.

Wenham, C. 2021. *Feminist Global Health Security*. Oxford: Oxford University Press.

Wesolowski, A. et al. 2012. "Quantifying the impact of human mobility on malaria". *Science* 338: 267–70.

White, R. et al. (eds) 2017. *The Palgrave Handbook of Sociocultural Perspectives on Global Mental Health*. Basingstoke: Palgrave Macmillan.

WHO 2019. *Programme Budget 2020–2021*. Geneva: WHO.

WHO 2020. *Noncommunicable Diseases Progress Monitor 2020*. Geneva: WHO.

WHO 2022a. "Noncommunicable diseases". 16 September. www.who.int/news-room/fact-sheets/detail/noncommunicable-diseases.

WHO 2022b. *World Malaria Report 2020: 20 Years of Global Progress and Challenges*. Geneva: WHO.

WHO n.d.a. "Measles-containing-vaccine first-dose (MCV1) immunization coverage among 1-year-olds (%)". Global Health Observatory. www.who.int/data/gho/data/indicators/indicator-details/GHO/measles-containing-vaccine-first-dose-(mcv1)-immunization-coverage-among-1-year-olds-(-).

WHO n.d.b. "WHO called to return to the Declaration of Alma-Ata". www.who.int/teams/social-determinants-of-health/declaration-of-alma-ata.
WHO n.d.c. "Health security". www.who.int/health-topics/health-security.
WHO n.d.d. "Constitution". www.who.int/about/governance/constitution.
WID 2020. "Global inequality data: 2020 update". 10 November. https://wid.world/news-article/2020-regional-updates.
Wigley, A. *et al.* 2020. "Measuring the availability and geographical accessibility of maternal health services across sub-Saharan Africa". *BMC Medicine* 18 (1). DOI: 10.1186/s12916-020-01707-6.
Wilkinson, R. & K. Pickett 2010. *The Spirit Level: Why Equality Is Better for Everyone*. London: Penguin Books.
Williams, A. & I. Luginaah (eds) 2022. *Geography, Health and Sustainability: Gender Matters Globally*. Abingdon: Routledge.
Workman, C. *et al.* 2021. "Understanding biopsychosocial health outcomes of syndemic water and food insecurity: applications for global health". *Journal of American Tropical Medicine and Hygiene* 104 (1): 8–11.
World Bank 1993. *World Development Report 1993: Investing in Health*. Washington, DC: World Bank.
World Bank 2020. "As Haiti braces for the COVID-19 pandemic, water, sanitation, and hygiene are more important than ever". 29 May. www.worldbank.org/en/news/feature/2020/05/29/as-haiti-braces-for-the-covid-19-pandemic-water-sanitation-and-hygiene-are-more-important-than-ever.
World Bank 2022. "Measuring poverty". 30 November. www.worldbank.org/en/topic/measuringpoverty.
World Bank n.d.a. "Health". www.worldbank.org/en/topic/health.
World Bank n.d.b. "What is food security?". www.worldbank.org/en/topic/agriculture/brief/food-security-update/what-is-food-security#:~:text=Based%20on%20the%201996%20World,an%20active%20and%20healthy%20life.
World Population Review n.d. "Country rankings: poverty rates by country". worldpopulationreview.com/country-rankings/poverty-rate-by-country.
Worobey, M. *et al.* 2022. "The Huanan seafood wholesale market in Wuhan was the early epicenter of the COVID-19 pandemic". *Science* 377: 951–9.
WTO 2021. "TRIPS and public health". www.wto.org/english/tratop_e/trips_e/pharmpatent_e.htm.
WTO n.d.a. "Who we are". www.wto.org/english/thewto_e/whatis_e/who_we_are_e.htm#:~:text=The%20overall%20objective%20of%20the,countries%20build%20their%20trade%20capacity.
WTO n.d.b. "TRIPS – trade-related aspects of intellectual property rights". www.wto.org/english/tratop_e/trips_e/trips_e.htm.
Xiao, X. *et al.* 2020. "Distribution and health risk assessment of potentially toxic elements in soils around coal industrial areas: a global meta-analysis". *Science of the Total Environment* 713. DOI: 10.1016/j.scitotenv.2019.135292.
Yamey, G. *et al.* 2022. "It is not too late to achieve global COVID-19 vaccine equity". *British Medical Journal* 376. DOI: 10.1136/bmj-2022-070650.
Yang, J., H. Mamudu & R. John 2018. "Incorporating a structural approach to reducing the burden of non-communicable diseases". *Globalization and Health* 14. DOI: 10.1186/s12992-018-0380-7.
Yearley, S. 2000. "Environmental issues and the compression of the globe". In *The Global Transformations Reader: An Introduction to the Globalization Debate*, D. Held & A. McGrew (eds), 145–60. Cambridge: Polity Press.
Yin, P. *et al.* 2020. "The effect of air pollution on deaths, disease burden, and life expectancy across China and its provinces, 1990–2017: an analysis for the Global Burden of Disease study 2017". *Lancet Planetary Health* 4 (9): e386–e398.

REFERENCES

Youde, J. 2020. "Philanthropy and global health". In *The Oxford Handbook of Global Health Politics*, C. McInnes, K. Lee & J. Youde (eds), 409–26. Oxford: Oxford University Press.

Young, I. 1990. *Justice and the Politics of Difference*. Princeton, NJ: Princeton University Press.

Yourkavitch, J. *et al*. 2018. "Using geographical analysis to identify child health inequality in sub-Saharan Africa". *PLoS ONE* 13 (8). DOI: 10.1371/journal.pone.0201870.

Yuen, A., A. Warsame & F. Checchi 2022. "Exploring the temporal patterns and crisis-related risk factors for population displacement in Somalia (2016–2018)". *Journal of Migration and Health* 5. DOI: 10.1016/j.jmh.2022.100095.

Zhao, C. *et al*. 2021. "The evolutionary trend and impact of global plastic waste trade network". *Sustainability* 13 (7). DOI: 10.3390/su13073662.

Zhou, M. *et al*. 2019. "Mortality, morbidity and risk factors in China and its provinces, 1990–2017: a systematic analysis for the Global Burden of Disease study 2017". *The Lancet* 394: 1145–58.

Zietz, S. & M. Das 2018. "'Nobody teases good girls': a qualitative study on perceptions of sexual harassment among young men in a slum of Mumbai". *Global Public Health* 13 (9): 1229–40.

INDEX

Aboriginal populations *see* Indigenous populations
access to healthcare 31, 33, 36, 39, 45, 50–2, 62, 66–70, 78–9, 81, 100–1, 107, 143–4, 162, 191
Afghanistan 51–2, 81, 95, 102, 104–5, 127
African Coalition for Epidemic Research, Response, and Training 87
Agbogbloshie scrap market, Accra, Ghana 124–5
agriculture 23, 32, 111, 119, 155–6, 163, 166, 180–1, 185
AIDS 2, 52, 66, 90
air travel, international 90, 151, 153, 163, 165
alcohol consumption 3, 46–8, 53, 58
Alma Ata declaration 75, 86
Amazonia 32, 119, 159, 169
American Indian and Alaska Native (AIAN) populations 53
Angola 28, 29, 178
antidepressants 61
antimicrobial resistance (AMR) 87, 165–6
aquaculture 6–7
Argentina 177
armed conflict *see* military conflict
arms transfers 125–6
arsenic, contamination of water 10, 116, 146–8
artisanal mining 117, 159
assemblage 8–10, 13, 97, 155, 158, 160, 165
asthma 125, 137
Aswan Dam, Egypt 8
asylum 94–5, 99, 102, 105, 110
austerity 11, 20–1, 30–1, 170
avian influenza (H5N1) 90, 163

Bangladesh 10, 59, 101–2, 146–8, 180
Basel Convention on waste 121
bauxite 162, 196
Bhopal, India 132–4
bilharzia *see* schistosomiasis

Bill & Melinda Gates Foundation (BMGF) 24, 26, 74, 76, 81, 84–6
"bird flu" *see* avian influenza
birth defects 33, 117–18, 124, 127
births 28–9, 33, 67
blood lead levels 119, 122, 124
borders 97–9, 160–2, 190
Borneo 153
Botswana 66, 70
Brandt Line 5
Brazil 10, 29–32, 50, 52, 57, 67–8, 156, 159, 168–9, 170, 177, 195
 Amazonia 31–2, 119, 159, 169
 Bolsonaro presidency 30–1
 healthcare system 30
 income inequality 25
 regional health inequalities 29–32, 52
British Columbia, Canada 6–7
Burkina Faso 28, 70, 176
Burundi 180

Cabo Delgado province, Mozambique 38–40, 69, 116–17
cadmium 116, 118, 124
"Calais Jungle" 102–3
Cambodia 111, 139, 143–4, 160, 180
Cameroon 28, 78, 109, 145
Canada 6–7, 112, 116, 126, 155, 164
cancer 116, 118–19
cardiovascular disease 46, 80, 119, 135, 177
Central African Republic 28, 60, 70, 180
Chad 29, 180
chikungunya 156, 179
child health 27, 33, 62, 77, 128, 143
Chile 139
China 35–7, 49–50, 59, 117–18, 121, 123–4, 126, 134, 136, 139, 153, 164, 171, 177, 194

INDEX

Belt and Road initiative 37
Covid-19 37, 50, 58, 86, 98, 103, 166–7, 180
income inequality 35
life expectancy 36–7, 49
regional inequalities 35–6
cholera 40, 144–5, 180
chronic obstructive pulmonary disease (COPD) 137
cigarettes *see* smoking
civets 164
climate change 21, 175–88
adaptation to 186–7
colonialism 176
food security 180–2
health impacts 176–80
and infectious disease 177–80
population displacement 184–6
coal mining 118–19
cobalt mining 117
Colombia 63–4, 95–6, 109, 156, 169
colonialism 6, 146, 155, 171, 176
commercial determinants of health 20–2
COVAX 172
Covid-19 3, 12, 24, 40, 68, 90, 91, 151, 166–72
incidence 169–70
mortality 169
vaccination 29, 40, 79, 170–2
crop burning 137
cyclones 40, 111

dams 8, 100, 141, 154, 195
decolonization 14–15, 27
deforestation 31, 154, 159, 175, 179
Delhi, India 56, 58, 64
Democratic Republic of the Congo 40–1, 52, 70, 100, 127, 180
colonial history 40
mineral resources 117
regional inequalities 41, 52
dengue 153, 156–9, 179–80
Denmark 109
depression 63–4 *see also* mental health
diabetes 25, 64
diamond mining 117, 162, 196
diarrhoea, in children 35, 122, 143–4, 163
diet 46, 56, 80, 180
dioxins 124
Disability-adjusted life years (DALYs) 159–60
disease ecology 2, 4–10, 13, 179
domestic workers 107–8
drinking water quality 40, 138
drought 177, 179, 184

earthquakes 111–12, 144
Ebola 2, 12, 83–4, 151, 160–2, 196
e-cigarettes 46, 48
e-waste 123–6
ecosyndemics 155, 169, 195
Ecuador 119, 157, 169, 183
Egypt 8, 46
emotions 8, 10, 106
English Channel, populations crossing 97–9
environmental justice 115–16, 131, 186
Eritrea 99, 102
Ethiopia 28–9, 70, 100, 106–7, 109, 127, 156
ethnography 2, 16, 45, 126
Every Child Protected Against Trafficking (ECPAT) 110
extensively drug-resistant tuberculosis (XDR-TB) 165

First Nations population groups, in Canada 6–7, 54–5
fishing 6–7, 39, 141, 180
flooding 111, 177, 180, 190
food insecurity 142, 180–4
forced displacement *see* population displacement

Gabon 67
Gambia 65, 181
Gates Foundation *see* Bill & Melinda Gates Foundation
Gaza Strip *see* Palestine
gender inequality 23, 50–2, 155, 182, 194
geographic information systems (GISs) 2, 16, 64, 69, 84, 152–3
geography of health *see* health geography
Germany 95, 102
Ghana 29, 66, 124–5, 139, 153, 160, 180
Gini index 25, 36
global environmental justice 115–16, 187
Global Alliance for Vaccines (GAVI) 80–2
Global Burden of Disease (GBD) programme 26–7, 42, 59, 136
Global Fund to Fight Aids, Tuberculosis and Malaria 80–1
global health governance 73–92, 189–91
global health security 87–91, 189–91
Global Health Security Index 90, 190
Global Outbreak Alert and Response Network (GOARN) 89
global positioning systems (GPS) 84, 153
Global Public Health Intelligence Network (GPHIN) 89
globalization 10–12, 36, 61, 74, 93, 163–4, 187, 195–7
Green Revolution 133

232

green spaces 63–4
Guatemala 65
Guinea 83, 160–2
Guyana 64

Haiti 2–4, 111–12, 144–5, 192
happiness *see* subjective wellbeing
Happy Planet Index 191
Health Alliance International 82–3
health geography 1–4, 9
health justice 192–3
heatwaves 175
historical trauma 53–4, 100
HIV 2, 52, 66, 81, 85, 90, 152, 155, 169
Hong Kong 137, 164
"horizontal" healthcare 33, 75, 85
horseshoe bats 164, 166
human rights 92, 147, 192–4
human trafficking 94, 108–11
humanitarianism 104
hunger 81, 90, 116, 180–2

illiteracy 31–4, 75
immunization 33, 51, 68–9, 75, 81, 101, 156
income inequality 25–6, 35, 63
India 24, 32–5, 51, 56, 61, 69, 81, 85, 100, 118, 123, 126, 132–7, 144, 168, 171, 175–7, 179, 186
  British rule 32–3
  regional inequalities 33–4, 51
Indigenous populations 6–7, 14, 32, 52–5, 66, 100, 116, 118, 139, 155, 169, 182–4, 192, *see also* Native Americans
Indonesia 46, 111–12, 122–3, 177
influenza 152, 163
informal settlements 55–8
Intergovernmental Panel on Climate Change (IPCC) 175, 182, 186
internally-displaced persons (IDPs) 95–7
International Health Regulations 88
International Monetary Fund (IMF) 21, 77, 196
International Organization for Migration 108
inverse-care law 66–7
Iran 60, 102
Iraq 100, 102, 106, 119–20, 127
Israel 100, 172
Italy 102

Kenya 28, 70, 111, 142, 153, 178, 180
Kerala, India 33, 68, 171
Kuwait 120

Lao PDR (Laos) 139, 159
latex 162

lead, in batteries 124, *see also* blood lead levels
Lebanon 100, 107, 156
leishmaniasis 154–6
Lesotho 46, 60
Liberia 83, 145–6, 160–2
Libya 98, 102–3
life expectancy 26, 31, 33, 36–7, 40, 54, 66, 127
liquid petroleum gas (LPG) 137
Lyme disease 8
lymphatic filariasis 154

Madagascar 29, 176, 180
malaria 2, 6, 13–14, 40–1, 74, 80–1, 156, 159, 177–8
Malaysia 6, 123, 159
Malawi 28, 70, 138, 142
Maldives 186
malnutrition 21, 33, 41, 49, 78, 90, 142, 181
maternal health 50–2
Mauritania 28
measles 29
Médecins Sans Frontières (MSF) 81–2, 84, 161, 172
medical anthropology 2, 154
medical geography 1, *see also* health geography
medical "tourists" 165
Mediterranean 46
Mekong river basin 139–41
mental health 53, 59–64, 101, 148, 182–4
Mexico 97, 121, 163, 168–9, 177
microcephaly 156 *see also* Zika virus
microplastics 122
Middle East Consortium for Infectious Disease (MECIDS) 87
migrant domestic workers 107–8, 169
migration *see* borders; mobilities; population displacement; refugees
military conflict 62, 156, 159
mineral extraction 39, 159–60
mobile phones, use in disease monitoring and healthcare 69–70, 153, 167
mobilities 93–4, 111–12
Mongolia 69
more-than-human health geography 9, 13
mortality
  adolescent 25, 27
  all-cause 25
  child and infant 25–7, 33, 37, 39, 50, 67, 78, 127
  heat-related 176–7
  maternal 51–2, 69, 73, 77–8
  premature 46, 134–5
mosquitoes 6, 10, 153–60, 177–9

INDEX

Mozambique 38–40, 50, 69, 81, 83, 158
  colonial history 38–9
  mineral resources 40, 116–17
  regional inequalities 39
multi-drug resistance tuberculosis (MDRTB) 165
Mumbai, India 32, 56, 168
Myanmar 29, 100–2, 139, 160

Namibia 29, 178
nanoplastics 122
Native Americans 52–3, 118, 140, 169
natural hazards and population displacement 111–12
neglected tropical diseases (NTDs) 145, see also lymphatic filariasis; onchocerciasis; schistosomiasis
neoliberalism 23, 75, 77, 85, 175, 192
nickel 116, 118
Niger 28
Nigeria 28–9, 109, 120, 146, 155–6
non-communicable diseases (NCDs) 12, 25, 41, 46–7, 49–50, 80, 103, 169, 181
non-governmental organizations (NGOs) 31, 74, 82–4, 86
non-representational theory 8
Norwegian Refugee Council 100, 186
nutrition 23, 27, 49, 78, 80, 103, 137, 165, 180–2

obesity 49, 80, 108
oil extraction and refining 116, 119–20
onchocerciasis 145–6, 154
One Health 13, 193–4
ontological security 105–7
oral rehydration therapy 75, 143

Pacific islands 186
Pakistan 61, 135, 175, 180, 190
Palestine 60, 62, 100, 103, 127, 172
pandemics 163, 181, 186, see also Covid-19
particulates 134–8
Partners in Health 82–3
People's Health Movement 86
Peru 119, 195
pesticides 132–3
pharmaceutical companies 61–2, 79, 81, 84, 171–2
philanthrocapitalism 74, 85
philanthropy 84–8
Philippines 108, 111, 177
Physicians for Human Rights 51
plague 10
planetary justice 192–3
pluriverse 15
poliomyelitis 156

political
  determinants of health 20–22
  ecology 4, 6–10, 66, 137, 140, 147, 151, 154, 156–60, 181, 186, 194–5
  economy 2–3, 6, 19, 134
pollution
  air 116, 131–8, 169
  "halo" 122, 132
  "haven" 121, 132
  transboundary air 134–5
  water 118, 139
population displacement 95–107, see also internally-displaced persons
  due to climate change 184–6
  due to conflict 41, 156
  due to infrastructure projects 100
  due to natural disasters 39, 111–12
Portugal, colonization of Mozambique 38–9
poststructuralist approaches in health geography 8–10, 22
post-traumatic stress disorder 62, 111
post-Westphalian governance 88–9, 139
President's Emergency Program for AIDS Relief (PEPFAR) 81
primary care 30, 67, 69, 75, 81
"Product Red" 86
public–private partnerships 74, 80–2

quarantine 10

racism 67, 105, 117, 162
refugee camps 102–4
refugees 93–107, 160
reproductive, maternal, new-born and child health (RMNCH) 27
resource extraction 39, 159–60
respiratory disease 133–8
risk 11, see also world risk society
river blindness see onchocerciasis
Rockefeller Foundation 84
Rohingya people 100–2
rubies, mining of 116–17
rural health and healthcare 28–9, 53, 69, 78, 95, 100, 137, 144–6, 155–7, 159, 163, 181–4
Russia 110, 126, 171, 190
Rwanda 28, 40, 67, 98, 1217, 144, 176, 180

sanitation 40, 55, 75, 86, 112, 138, 142, 159
São Paulo, Brazil 57
SARS-CoV-2 see Covid-19
Saudi Arabia 126
schistosomiasis 8, 145–6, 154, 179–80
science and technology studies (STS) 9
sea level rise 181, 186, 190
security see global health security

self-reported health 25, 31, 53, 100
Senegal 170, 194
severe acute respiratory syndrome (SARS-CoV-1) 90, 151, 163–5
sex ratio 51
sex trafficking 108–11
Sierra Leone 14, 28, 83, 106, 127, 160–2, 180
Singapore 107–8, 164, 169
situated knowledge 15
slavery 108–11
sleeping sickness *see* onchocerciasis
slums *see* informal settlements
smallpox 10, 74
smoking 21, 46–8
snail fever *see* schistosomiasis
social determinants of health 20, 60
social movements 86–7
Somalia 90, 105, 185–6
South Africa 3, 24, 49, 63, 67
South Korea 134
South Sudan 95–6, 144, 156, 180
space–time convergence 10, 152
Spain 97
spatial diffusion 159, 167, 170
stigma 55, 58, 163
Stockholm International Peace Research Institute 126
structural adjustment policies (SAPs) 39, 75, 77–8
stunting, in children 33, 56
sub-Saharan Africa 24, 26–7, 55, 67, 78, 87, 97, 120, 145–6, 178–80
Sudan 70, 95–6, 98, 102, 106, 144, 156
Sustainable Development Goals (SDGs) 22–4, 28, 34, 36, 46, 49, 59, 65, 80, 107–8, 135–6, 138, 145, 180, 187, 191–3
swine flu (H1N1) 163
syndemics 142, 154–5, 169, 181, 194–5
Syria 21, 62, 90, 95–6, 100, 102, 106, 156, 180, 184

Thailand 110–11, 123, 137, 139, 160
therapeutic landscapes 3, 61, 65
time–space compression/convergence 10, 152
titanium 162, 196
tobacco 21, 46, 79, 103, 196
Trade-Related Aspects of Intellectual Property Rights (TRIPS) 79
transboundary
  air pollution 134–5
  water flow 138–42
trauma 53–4, 100
tsunami 111
tuberculosis (TB) 34, 41, 80, 165, *see also* extensively drug-resistant tuberculosis; multi-drug resistance tuberculosis

tubewells 146–7
Turkey 21, 95, 102, 123

Uganda 146, 178, 180, 183
Uighur population, China 37
Ukraine 93, 172, 190
United Nations (UN) 50, 138, 144, 186
United Nations Children's Fund (UNICEF) 62
United Nations High Commissioner for Refugees (UNHCR) 95, 184
universal truths 14–15
uranium mining 118
urban health 3, 28–9, 53, 55–9, 95, 137, 158–9, 170, 176, 184
Uruguay 170
Uttar Pradesh, India 68

vaccination 37, 51, 81, 170–2, *see also* Covid-19; immunization
vaccine nationalism 171
vaping *see* e-cigarettes
Venezuela 95, 156, 159
"vertical" healthcare 33, 75, 85
victim-blaming 120, 138
Vietnam 110, 123, 139, 144, 164, 170
violence
  domestic 52
  epistemic 125
  sexual 41, 52, 57, 100, 108, 110, 144
  slow 124–6
  structural 2–3, 51, 54, 99, 102, 125–6, 138, 155, 159, 169, 195

walkability 49
WaSH (water, sanitation and hygiene) 20, 143–4, 180
waste, *see also* e-waste
  hazardous 120–6
  movement of 120–6
  plastic 122
water
  access to clean drinking 56, 143–8
  pollution 118, 139
  quality of 112, 143–8
  security of 138–42, 146–8
Washington Consensus 77
wellbeing 4, 64–6
  subjective 4, 65, 72, 191–2
Westphalian governance 88–9, 139
wet markets 164–6
World Bank 24, 26, 46, 74, 77–8, 120, 190, 196
World Economic Forum 86–7, 91
World Happiness Report 65
World Health Network 86

World Health Organization (WHO) 26–7, 37, 46, 48, 74–7, 81, 87–9, 162, 193
World Inequalities Database 24
World Social Forum 86
World Trade Organization (WTO) 78–9
world risk society 11, 121, 187
WorldPop team (Southampton) 152–3
Wuhan, China 166–7, 196

Yemen 50, 90, 126–7, 180

Zambia 67, 69, 180–1
Zika virus 10, 156–7
Zimbabwe 143
zoonotic disease 166, 177, 179